ENVIRONMENT AND DEVELOPMENT TRAJECTORY: A FRESH LOOK ON THEORY AND EMPIRICS

ENVIRONMENT AND DEVELOPMENT TRAJECTORY: A FRESH LOOK ON THEORY AND EMPIRICS

SOUMYANANDA DINDA

Nova Science Publishers, Inc.
New York

For permission to use material from this book please contact us:
Telephone 631-231-7269; Fax 631-231-8175
Web Site: http://www.novapublishers.com

LIBRARY OF CONGRESS CATALOGING-IN-PUBLICATION DATA

Dinda, Soumyananda.
Environment and development trajectory : a fresh look on theory and empirics / Soumyananda Dinda.
p. cm.
Includes bibliographical references and index.
ISBN 978-1-60692-246-0 (softcover : alk. paper)
1. Economic development--Environmental aspects. 2. Environmental economics. I. Title.
HD75.6.D56 2009
333.7--dc22

2008041202

Published by Nova Science Publishers, Inc. ✝ New York

CONTENTS

PREFACE

The last few years have witnessed raising concerns about whether environmental constraints will limit development or whether development will cause serious environmental damage – in turn impairing the quality of life of this and future generations. This book attempts to explore the causal relationship between economic development and environmental quality change. A theoretical base is developed to explain the environment-development relationship in the endogenous growth framework. This book also incorporates few empirical studies and attempts to integrate environmental considerations into developmental policymaking.

This book contains ten chapters. Chapter 1 introduces the issues that have been covered in this book. Chapter 2 presents a brief survey of the relevant literature. Chapter 3 develops a theoretical base for the empirical observations. Chapter 4 reports on an empirical study on the relationship between environmental quality and economic growth. Chapter 5 examines the issue of causality of the relationship between income and emission of CO_2 in a country group-specific analysis based on a panel data set. Chapter 6 presents a further analysis of the nature of cointegration between income and emission using the technique of cointegration analysis for panel data set. Chapter 7 provides a study of the pattern of distribution of income and emission and their interrelationship. Chapter 8 examines the impact of globalization on environment. Chapter 9 explains the technological evolution in the process of economic development. Finally chapter 10 concludes with future direction of research.

This book is prepared for advanced graduate students, policymakers and professionals working on development and environmental economics.

I am grateful and indebted to Professor Dipankor Coondoo for suggestion, comments and collaborative works. Without his kind guidance this work could never have been completed. This, therefore, is a nice opportunity to express my sincerest gratitude to him for his excellent academic guidance at every stage of my research career. I gratefully acknowledge the help provided by Professor Manoranjan Pal.

I am grateful to the reviewers who helped in bringing this book in its present form by offering valuable comments and suggestions on draft version of this book.

I acknowledge the Department of Economics and International Development, University of Bath, UK for giving me opportunity to complete this work. I also acknowledge S. R. Fatepuria College, Beldanga, Murshidabad for granting leave (lien) to complete this work at Bath, UK.

I would like to thank the seminar participants at Indian Statistical Institute, Madras

School of Economics, Delhi School of Economics, Institute of Economic Growth, Jadavpur University, Oldenburg University, Birmingham University, Reading University and University of Bath where parts of this book were read out.

I could not have done this by myself and I totally appreciate the supports and efforts of all who were involved in making this book possible, which includes the referees, Chief Editor of Nova Science Publication, editorial staff of Nova Science Inc. The author gratefully acknowledges the assistance provided by Chief Editor and other editorial staff for this book.

I must heartily thank my friends, Arijit, Atanuda, Brati, Diganta, Rajudi, Manishadi, Samarjit, Monipuspak, Udaybhanu, Debabrata, Sonali, Bidisha, Rituparna, Debashis, Anup and Shahana for their constant encouragement.

It is needless to acknowledge the most needed support and encouragement given to me by my family members especially eldest brother Panchananda, Boudi, and other relatives, especially Anilda. My love and gratitude goes to my mother and to Moni, Bappa, and Bittu for their perseverance and finally to Jayashri for 'being there' all through.

Soumyananda Dinda
Bath, UK,. April 2008

Chapter 1

INTRODUCTION

This book explores some issues of environmental economics arises from development activities. A concept crystallizing in the development and environmental economics literature is the notion that socioeconomic and environmental measures follow predictable paths associated with growing per capita income. In this context, the nature of relationship between economic development and environmental quality has become the focus of increasing attention. The issue of whether environmental degradation increases monotonically, decreases monotonically, or at first increases and then declines along a country's development path, has critical implications for policy. Whilst development through industrialization brings higher incomes and well-being, this seems to act as a *magnifier* of environmental degradation. On the other hand, growing environmentalism is perceived to act as an impediment to economic development. Economic developments through rapid industrialization and growing environmental consciousness together have generated a heated debate on how economic development may be linked with environment.

The linkage of environmental quality with economic development evoked much discussion in the last decade (i.e., 1990s). The World Development Report (World Bank 1992) presented cross-sectional evidences on the relationship between different indicators of environmental quality and per capita national income across countries. Other studies (e.g., Grossman and Krueger, 1991, 1995; Shafik, 1994; Selden and Song, 1994; McConnell, 1997; de Bruyn et al., 1998; Rothman, 1998; Suri and Chapman, 1998; Stern and Common, 2001; Stern, 2002) documented an inverted U-shaped relationship between environmental degradation and income. The common point of all these studies is the assertion that environmental degradation increases initially, reaches a maximum level and after that declines as an economy develops. This systematic inverted-U relationship has been called as the *Environmental Kuznets Curve* (*EKC*) following the work of Kuznets (1955), who postulated a similar relationship between income inequality and economic development.

The *EKC* relates to the issue of the impacts of economic development on environment. To understand why and how economic developmental issues get linked to concerns about environmental degradation requires a careful study. In fact, detailed studies are needed to understand the specific nature and the shape of *EKC*. Re-examination of the relationship between environmental quality and economic growth thus remains an open issue.

The *EKC* results suggest that economic growth could be compatible with environmental improvement provided appropriate policies are taken. On the other hand, effective environmental policies may be implemented when income grows. However, before adopting a policy, it is necessary to understand the exact nature of the causal relationship between economic growth/development and environmental quality. The question therefore is *whether economic growth can be a part of the solution rather than the cause of environmental problem*. This is indeed the primary motivation for this study. Here empirical evidences of the link between income and environmental degradation have been searched while the desirability of development is universally recognized, recent years have witnessed rising concern about whether environmental constraints will limit development or whether the ongoing process of development will cause serious environmental damage. Thus, the causal relationship between economic activity (viz., consumption and production) and environmental quality deserves to be explored carefully to bring out explicitly the linking economic development/growth to environmental quality change.

Torras and Boyce (1998) have brought in the distributional issue in the discussion of income-emission relationship[1]. The inequality of the income distribution may be an important determinant of the extent of environmental degradation and more specifically, income redistribution may reduce the inequality by affecting the society's demand for better environment. Knowledge of the pattern of distribution of emissions and intertemporal choice may also help to formulate appropriate environmental policy. All of these issues naturally call for formal analyses and empirical verification. The issues that seem to be central to a comprehensive examination of economic development – environmental quality relationship are:

1. Does the inverted-U relationship between pollution and income (*EKC*) exist?
2. Is there any causal mechanism between income and environmental variables?
3. What is the role of other factors such as income distribution? To be specific, how does income distribution affect the global emissions through distribution of emissions?
4. What is the impact of globalization on environment?
5. Does technological progress improve our environment in very long run?

The aim of this book is to explore the above mentioned issues in detail. Briefly, there are five major parts of this study. The first part examines whether for air pollutants like *SPM* and SO_2 the *EKC* hypothesis is supported by the available data set(s), analyses and provides explanation of the empirical results obtained. The second part investigates the nature and direction of causality between CO_2 emission and income growth implicit in a cross-country panel data set, using appropriate econometric tools and explains the observed patterns of causality. The third part tries to bring out the pattern of relationship that exists between the cross-country distributions of income and emission and its shifts over time. Fourth part highlights the debate of globalization and its impact on environment. Finally, this book focuses on how technological progress really improves our environment in long run. The results are hoped to give valuable insights into the important and growing concern about interrelationship between environmental changes and economic development.

---

[1] In the discussions that follow, the phrases like income-emission relationship, economic development-environmental quality relationship etc. will be used interchangeably.

1.2. The Plan of this Book

The plan of this book is as follows: Chapter 2 provides a brief review of the relevant literature up to date.

Chapter 3 develops a theoretical model in endogenous growth framework. The model provides the explanation for different trajectory for various economies, under certain conditions. Considering a closed economy, one part of capital is used for commodity production, which generates pollution that degrades existing environment, and the remaining part is used for abating pollution or upgrading environment. Sufficient abatement activity improves or restores environmental quality. This sufficient abatement activity associated with commodity production could only lead optimally towards steady state. In the economy, allocation of capital for abatement activity varies over time. Thus, a change from insufficient to sufficient allocation of capital for abatement activity is the basis for an inverted-U relationship between environmental quality and economic growth.

Chapter 4 re-examines the *EKC* hypothesis using two important ambient air quality indicators, viz., *SPM* and SO_2. The results do not support the *EKC* hypothesis. In contrast, for SO_2 an inverse relationship with income per capita is obtained, while for *SPM* a U-shaped, rather than an inverted-U, relationship is found. The estimated curve turns upward around a per capita annual income level of $ 12500, which represents a rather high level of material consumption that appears to be unsustainable, given the state of available technology. Further rise of income beyond the threshold level can support consumption only at the cost of a slow but steady deterioration of the environmental quality, which may slow down the economic growth. Here, the association between the level of economic activity and environmental degradation has been examined without explicitly discussing the nature of causation between them. The pattern of causal relationship between economic growth and environmental quality has been explored in the next chapter.

Chapter 5 presents a detailed examination of the causality aspect of the income-environment relationship. Customarily, a unidirectional causality with income causing environmental changes and not vice versa is presumed. The validity of this presumption is now being questioned. This chapter analyses the results of a study of income-CO_2 emission causality based on Granger causality technique. Briefly, the results indicate three different types of causal relationship holding for different country groups. Briefly, for the developed country groups of North America and Western Europe (and also for Eastern Europe), the causality is found to run from emission to income. For the country groups of Central and South America and Oceania, causality from income to emission is obtained. Finally, for the country groups of Asia and Africa, the causality is found to be bi-directional. The estimated regression equations further suggest that for the country groups of North America and Western Europe the growth rate of emission has become stationary around a zero mean, and a shock in the growth rate of emission would generate a corresponding shock in the growth rate of income. In contrast, for the country group of Central and South America and Oceania a similar shock in the income growth rate is likely to result in a corresponding shock in the growth rate of emission. Finally, for the county groups of Asia and Africa, the income and the emission growth rates seem to reinforce each other. These results naturally suggest that a further exploration of the patterns of underlying long run equilibrium relationship for the sake of completeness of this study. This has been done in the next chapter, using the technique of cointegration analyses.

Chapter 6 presents the results of cointegration analyses and related error correction model (*ECM*) estimation. Applying the panel unit root test procedure of Im et al., (2003) (*IPS*), the unit root hypothesis has been examined for the time series processes of the concerned variables. The panel data cointegration test based on Engle-Granger Cointegration technique has also been performed and the *ECM* has been estimated to explore the nature of dynamics implicit in the given panel data set. The cointegrating relationship is found to exist for some country groups implying thereby that over a long period of time income and emission tends to move in unison.

Chapter 7 explores the relationship between the patterns of cross-country distribution of income and CO_2 emission and temporal shifts in such a relationship. Here environmental quality demanded is treated as a *private good,* not a public good as done in other studies. The technique of Lorenz and specific concentration curve analysis are used as the basic analytical framework to argue that a measure of relative distributional inequality of income should be used as an explanatory variable in the *EKC* relationship along with the mean income level. The empirical results confirm that the pattern of cross-country income distribution has significant effect on the mean level of emission for all the country-groups considered.

Chapter 8 examines the impact of globalization on environment. This chapter examines empirically how trade affects the environment that is crucial for resolving current policy debates. Applying panel data technique we examine the impacts of globalization on pollution level, pollution intensity and relative change of pollution for the developed (OECD) and developing (Non-OECD) country groups and the world as a whole. This chapter examines the factor endowment and pollution haven hypotheses that predict how trade affects the environment. Interaction effects also play a crucial role to determine the impact of globalization on environment. In this study we use CO_2 emission and observe that the impact of globalization on environment heavily depends on the basic characteristics of a country and its dominating comparative advantage. The empirical results suggest that globalization increases CO_2 emission, which is the main culprit of the global warming.

Chapter 9 provides the intuition of technological evolution with development. Technological innovation increases productivity. It also generates few externalities which was unknown at the time of installation of the technology. As soon as externalities are identified, society formulates laws and regulates it. Thus, society creates the condition for new innovations. So, a cyclical pattern arises in which technology first diffuse, then become regulated and finally is phased out by another new technology.

Finally, Chapter 10 provides some concluding comments with future directions of research.

APPENDIX

There are two appendices to this book. Appendix A presents the list of country groups considered in most of empirical analysis reported in the book. Appendix B briefly presents the econometric methods used in this book –viz. the methodology of Granger Causality Test, Panel Unit Root and Cointegration Tests.

Chapter 2

ENVIRONMENTAL KUZNETS CURVE: A REVIEW

2.1. INTRODUCTION

The linkage between environmental quality and income has evoked considerable discussion since the pioneering work of Grossman and Krueger (1991). A sizeable literature that has emerged comprises theoretical and empirical studies concerning the nature of the income-environment relationship. This literature centers on the *Environmental Kuznets Curve* (*EKC*) hypothesis that postulates an inverted-U shaped relationship between environmental deterioration and economic growth. The logic of *EKC* hypothesis is intuitively appealing. It is postulated that environmental degradation rises rapidly in the early stage of economic growth primarily because the society attaches a higher priority to growth of material output and income over the demand for clean air and water (Dasgupta et al., (2002)). Since the awareness of environmental problem is low at this stage, environment-friendly production technologies are not available /used. Typically, income growth at this stage takes place through expansion of agriculture and other intensive resource extracting activities, thus causing an enormous pressure on environment and rising pollution. Beyond a threshold level of development, however, a high and rising income level induces people to value environmental quality more than before. This environmental awareness induces several qualitative changes in the economy in favour of less resource intensive production, stricter enforcement of environmental regulations and standards etc. In sum, thus, the story of *EKC* hypothesis is a narration of the process of evolution of a clean agrarian economy to a polluting industrial economy and ultimately to a clean service economy (Arrow et al., (1995)).

In this chapter a review of the literature on the study of *EKC* is presented. The chapter is organized as follows: the genesis of *EKC* hypothesis is briefly explained in section 2.2; various environmental quality indicators used in the *EKC* literature are discussed in section 2.3; a brief review of some specific explanation of *EKC* hypothesis with empirical evidences is presented in section 2.4; major theoretical analyses put forward in the context of *EKC* are reviewed in section 2.5 and finally, a conclusion is drawn in section 2.6.

2.2. GENESIS OF EKC

The origin of the *EKC* debate lies in the so-called growth controversy[2]. It is argued by many that economic growth degrades environmental quality. Beckerman (1992), in contrast, suggested that the economic growth by itself could be a panacea for environmental degradation. Supporting this view, Panayotou (1993) also claimed that economic growth could be a precondition for environmental improvement, particularly in developing countries. The essential argument underlying the Beckerman-Panayotou proposition is perhaps the following: if a country in its course of development is able to reach a high enough income stage, it will be willing and able to afford an income generation strategy leading to improved environment. This is in direct conflict with the pessimistic view of the Club of Rome – viz., global economic development will be unsustainable unless a zero growth rate steady state strategy of development is pursued. In fact, a bridge between these two opposite views about the future of the world ecology would be the idea of a so-called *development path* which provides a stage-based link between environment and economic growth (Selden and Song (1994), Grossman and Krueger (1995), Stern (1998, 2004)). Briefly, this notion of development path suggests that in the early stage(s) of economic growth a poor society will prefer and strive for a fast growth of output and income at the cost of extensive environmental degradation, whereas once a threshold level of development has been crossed, the same society will attach a much greater value to the environmental quality and hence take measures to ensure improvement of the environment. The *EKC* hypothesis is essentially an expression of this idea of a development path. Obviously, this hypothesis calls for an empirical verification.

From the beginning of 1990s empirical data on various pollutants have been made available by agencies like the Global Environmental Monitoring System (GEMS), the Oak Ridge National Laboratory (ORNL), World Resources Institute (WRI), International Energy Agency (IEA), United Nations Environment Programme (UNEP) etc. This data availability induced several authors to test the validity of *EKC* hypothesis. The first empirical study[3] appeared in the National Bureau of Economic Research (NBER) working paper by Grossman and Krueger (1991) and after that a number of studies followed. In 1990s and onwards, the Kuznets Curve[4] took on a new existence and has become a vehicle for describing the relationship between pollution and income.

---

[2] Before 1970 there was a belief that the consumption of raw materials, energy and natural resources would grow almost at the same rate (i.e., at the steady state rate) as an economy grew. In early 1970s the Club of Rome's *Limits to Growth* view (Meadows et al., (1972)) expressed concern for availability of natural resources of the Earth. Briefly, the environmental economists of the Club of Rome argued that the finiteness of environmental resources would prevent economic growth and therefore urged for a steady state with zero rate of growth to avoid dramatic ecological consequences in future.

[3] First set of empirical *EKC* studies appeared independently in three working papers: an *NBER* working paper as part of a study of the environmental impacts of *NAFTA* (Grossman and Krueger (1991)), the World Bank's (1992) World Development Report, and a Development Discussion paper as part of a study for the International Labour Organization (Panayotou (1993)). Grossman and Krueger (1991) in *NBER* working paper, which was later published in 1993, first pointed out an inverted-U relationship between pollutants and income per capita. Kuznets' name was attached to the inverted-U relationship between pollution and economic development later due to its resemblance to Kuznets' inverted-U relationship between income inequality and economic development.

[4] Kuznets (1955) predicted that the relationship between per capita income and income inequality would be an inverted U-shaped one. As per capita income increased, income inequality would also increase at first and then start declining beyond a turning point. That means, the distribution of income becomes more unequal in early

2.3. A Brief Review of EKC Analysis Based on Various Environmental Quality Indicators

In the absence of a single environmental indicator, it is possible to distinguish three main categories of environmental quality indicators that have been used in the empirical studies in the *EKC* literature – viz., air, water and other environmental quality indicators. Results of studies based on these indicators are briefly discussed below.

Air Quality Indicators

The measures of urban and local air quality[5] indicators generally show an inverted-U relationship with income. Several studies[6] confirm this, but there are a few empirical evidences[7] against the validity of the *EKC* hypothesis (Jha and Murthy (2003), Perman and Stern (2003), Stern (1998), de bruyn and Heintz (1998), Ekin (1998) and Cole (2003)). Generally, the literature does not find much evidence in support of the *EKC* hypothesis for air pollutants that have indirect or little impact on health. Both early and recent studies observe that CO_2 emission increases monotonically with rising income. It should be noted that most of the air pollutants are energy related such as SO_2, CO_2, CO and NO_x. Several authors (de Bruyn and Opschoor (1997), Sengupta (1997)) find evidence of N-shaped curve for a few indicators[8].

Water Quality Indicators

For water quality indicators, the empirical evidences on the validity or otherwise of the *EKC* hypothesis are mixed. Three main categories of indicators are generally used as measures of water quality - viz., concentration of pathogens in the water (fecal and total coliforms), amount of heavy metals and toxic chemicals (lead, cadmium, mercury, arsenic and nickel) discharge in water by human activities and measure of deterioration of the water oxygen regime (dissolved oxygen, biological and chemical oxygen demand i.e., *BOD* and *COD*). A few indicators provide evidence in support of the *EKC* hypothesis, but most of the studies have conflicting results about the shape and peak of the *EKC* (Shafik (1994), Hettige et al., (2000)).

---

stage of growth and later the distribution moves towards greater equality as economic growth continues. This relationship between income per capita and income inequality can be represented by a bell–shaped curve. This observed empirical phenomenon is popularly known as the Kuznets curve.

[5] Cole et al., (1997) observed that significant *EKC*s existed only for local air pollutants like SO_2, SPM, NO_x, and CO. Selden and Song (1994) focussed on urban air concentrations with a peak at lower income levels than total per capita emissions. In contrast, Horvath (1997) observed that the global environmental indicators like CO_2 emission, municipal waste and energy consumption either increased monotonically with income or else have high turning points with large standard errors (Holtz-Eakin and Selden (1995)).

[6] See, Grossman and Krueger (1995), Selden and Song (1994), Shukla and Parikh (1992), Roberts and Grimes (1997), Bradford et al., (2000), Halkos (2003), Jha (1996), Tucker (1995), Roca (2003), Hilton and Levinson (1998), Kahn (1998), Taskin and Zaim (2000), Unruh and Moomaw (1998) etc.

[7] Harbaugh et al., (2002) point out that there is little empirical support for an inverted U shaped relationship between income and several important air pollutants.

[8] The *EKC* may not hold in the long run (de Bruyn et al., (1998)). One may visualize an N-shaped curve, which exhibits the inverted U-curve initially but beyond a certain income level the relationship between environmental pressure and income again turns positive. There can be a secondary turning point (\$10,000 - \$15,000) at which the levels of ambient air pollution increase. This is the possibility of the *relinking hypothesis* (de Bruyn and Opschoor (1997), Sengupta (1997)).

Other Environmental Quality Indicators

Studies using other environmental indicators mostly do not support the *EKC* hypothesis. All the studies observe that environmental problems having direct impact on the population (viz., access to sanitation and safe drinking water) tend to improve steadily with economic growth. On the contrary, environmental problems having indirect impact on people (e.g., municipal solid wastes, CO_2 emission) have no tendency to decline. Finally, the evidences on the *EKC* relationship are highly conflicting in case of deforestation (Bhattarai and Hammig (2001), Bulte and Soest (2001), Koop and Tole (1999)).

A possible explanation of these conflicting results might be the application of different methodologies and evaluation of several environmental quality indicators using different cross-sectional cross-country data sets in various *EKC* studies. Therefore, to synthesize the *EKC* studies, in this context, Meta-analysis[9] could be an appropriate and useful technique that might confirm the *EKC* literature (Cavlovic et al., (2000)).

2.4. A Brief Review of Specific Explanations for the *EKC*

The *EKC* hypothesis actually summarizes an essentially dynamic process of change – viz. as income of an economy grows over time, initially emission level rises, reaches a peak and then starts declining after a threshold level of income has been crossed. However, the statement of the hypothesis makes no explicit reference to time. Truly, the *EKC* is a long run phenomenon; it describes a development trajectory for a single economy that grows through different stages over time. In their process of development individual countries[10] generate income and emission, which also follow one and the same *EKC*, *ceteris paribus*. Empirically this development trajectory can be observed in cross-country cross-sectional data, which represents countries belonging to different (low, middle and high) income groups corresponding to their emission levels. Assuming all countries follow one and the same *EKC*, at any point of time, it should be observed that poor countries are mostly at the rising part of *EKC*, developing countries are at the part of the *EKC* where it is approaching the peak or about to cross it and the rich countries are in the falling part of the *EKC*. Several factors can be responsible for shaping the *EKC*. In the following sub-sections, the potential partial effect(s) of these factors are analyzed briefly.

---

[9] A Meta-analysis is a statistical method of synthesizing results of similar empirical studies to determine whether credible conclusions about prior study results can be made.

[10] Vincent (1997) shows that pollution-income relationships from the cross-country studies fail to predict the trends in air and water pollution in Malaysia. In this context, Stern et al., (1996) point out that a more fruitful approach to the analysis of the *EKC* would be the examination of the historical experience of individual countries. Following qualitative historical analysis, Lindmark (2002) examines the inverted-U trajectory of Swedish CO_2 emissions during the period 1870-1997 and interprets the results within the context of growth regimes. Similarly, large differences in state level per capita emissions/pollutants are observed because of the enforcement of federal pollution laws in the USA (Carson et al., (1997), List and Gallet (1999) and Selden et al., (1999)).

2.4.1. Income Elastic Demand for Environmental Quality

The most common explanation of the shape of *EKC* is in terms of income elasticity of demand for environmental quality. As income grows, people achieve a higher standard of living and care more about the quality of environment they live in and demand improvement of it. Thus, the income elasticity of environmental quality demand is invoked in the literature as the main reason for the reduction of pollution (emission) level with rising income (Shafik (1994), Kristrom and Riera (1996), Carson et al., (1997), Komen et al (1997), Schmalensee et al.(1998) and McConnell (1997)). This implies that people attach increasing value to environmental amenities when a country achieves a sufficiently high income level (Selden and Song (1994)). This would be reflected through *defensive* expenditures, donations to environmental organizations and/or use of less environmentally damaging products and technologies. Magnani (2000) confirms it by examining OECD data on public R&D expenditure for environmental protection.

It should be mentioned that consumers with higher incomes may not only be willing to spend more for *green* products, but also create pressure for environmental protection regulations and institutional reforms such as promulgation of environmental legislation and creation of market-based incentives to reduce environmental degradation. Systematic efforts for reducing pollution in high-income countries is more likely to be observed if economic growth accompanies improvements in other social indicators, particularly income inequality[11] (Heerink et al., (2001)), education and information accessibility (Bimonte (2002)) that shift social preferences away from consumption of private goods toward public goods (viz., environmental amenities).

2.4.2. Scale, Technique and Composition Effects

Economic growth is thought to affect the environmental quality through three different channels –viz., scale, technique and composition effects (Grossman and Krueger (1991), Canas et al., (2003)). Growth exhibits scale effect on the environment. The increasing output requires more inputs and thus more natural resources are used up in production process, which generate more wastes and emissions as by-products. These in turn degrade environmental quality. Economic growth on the other hand may have a positive impact on the environment through a composition effect. As income grows, the structure of the economy may tend to change such that the share of cleaner activities in GDP is gradually raised. Environmental degradation may tend to increase as the structure of the economy changes from rural to urban or agricultural to industrial and then it may start to fall if another structural change takes place due to gradual replacement of energy intensive industrial activity by knowledge based technology-intensive industrial activity. Technological progress may also play a major role in this process of transformation to a cleaner environment by accelerating economic growth and at the same time by helping in the substitution of dirty and obsolete technologies by cleaner ones. This is the so-called *technique effect* of economic growth. The *EKC* suggests that the negative impact on the environment of the scale effect tends to prevail in the initial stages of growth, but eventually it gets outweighed by the

[11] It produces a gap between a country's willingness to pay and ability to pay for environmental protection.

positive impact of composition and technique effects that help to reduce emission level (Stern and Common (2001), Pasche (2002)).

2.4.2.1. Technological Progress

Generally, technological progress results in a greater efficiency in the use of energy and materials. In other words, with economic growth a given amount of goods can be produced with successively reduced burden on natural resources and environment. In this context, it should be mentioned that there is a growing trend among industries to reconsider their production processes and thereby take environmental consequences of production into account. This concerns not only traditional technological aspects of production but also organization of production and design of products. Lindmark (2002) observes that technological changes associated with the production process may also result in changes in the input mix of materials and fuels. In this context it may be mentioned that international trade generally facilitates diffusion of technology that prevents economic late-comers from requiring the same levels of materials and energy inputs per unit of GDP than what older industrialized countries needed in the past. This may allow developing countries to *dive through* the *EKC* (Magnani (2000)). Munasinghe (1999) suggests that developing countries may learn from the experiences of industrialized nations and restructure growth and development to *tunnel* through any potential *EKC* - thereby avoiding going through the same stages of growth that involve relatively high (and even irreversible) levels of environmental harm.

2.4.2.2. Structural Change

Along with economic development the societies advance with their social, legal, and fiscal infrastructures that are essential for enforcing environmental regulation[12]. It is true that institutional changes triggered by citizens' demand for cleaner environments are more likely to occur in democratic countries[13] (World Bank (1992)). These (the socio-economic changes) may lead to changes in the patterns of production. It is observed that generally the production structure shifts rapidly in industrializing and developing countries where as it remains more or less stable in developed countries. In a developing economy, the sectoral composition also changes rapidly changing thereby the industry's share in *GDP*. Such changing compositions of economic activity (in combination with trade), which is due to the shift in the structure of the economy (de bruyn (1997), Panayotou (1997, 1999), Kaufmann et al., (1998), Stern and Common (2001), Stern (2002)), may obviously have a major impact on environment. Some times an external shock may also force the structure of the economy to change. For example, oil shocks of the 1970 have caused an enormous structural economic transition world over towards environment-friendly technology that helped reduce emission (Moomaw and Unruh (1997), Unruh and Moomaw (1998)).

---

[12] In case of weak regulators, different social groups and local communities pursue informal regulation and often use other channels to induce pollution reduction by local factories in a process of `informal regulation' (Pargal and wheeler (1996), Dasgupta et al., (2001), Bhattarai and Hammig (2001)).

[13] However, opposite results can be found when the samples are divided into a subset of high and low income countries (Torras and Boyce (1998)). Most of the pollutants are substantially lower in more democratic low-income countries.

2.4.3. International Trade

It is believed that international trade is an important factor that helps to explain the *EKC*. Environmental quality could decline through the *scale effect* as increasing trade volume (especially export) would expand the size of the economy thereby increasing the extent of pollution. Thus, trade might be a cause of environmental degradation, *ceteris paribus*. As Antweiler et al., (2001) and Liddle (2001) argue, trade may be good for environment as well. Trade may improve the environmental quality through *composition effect* and *technological effect*. As income rises through trade, environmental regulation is tightened and as a result pollution reducing innovation gets promoted. The *composition effect* is attributed to two related hypotheses, viz., the *Displacement Hypothesis* and the *Pollution Haven Hypothesis*. Basically these two *hypotheses* are the same. Both hypotheses predict that polluting industries concentrate in developing countries with low environmental standards. It should be noted that the differences in the consumer preferences for a cleaner environment in rich and poor countries induce these hypotheses.

It is observed that changes in the structure of production in developed economies are not accompanied by equivalent changes in the structure of consumption. This could be explained by the *EKC,* which actually record the shifting of dirty industries to less developed economies. As Rothman (1998) speculates, what appears to be an improvement in environmental quality may in reality be an indicator of increased ability of consumers in wealthy nations to distance them from the environmental degradation associated with their consumption. The mechanisms through which such distancing takes place may include both moving sources of pollution away from the people and moving people away from pollution sources[14]. Thus, in general, the phenomenon of *distancing* may be a possible source of *EKC* results. Hettige et al (1992) observe that toxic intensity grew rapidly in high-income countries during the 1960s and this pattern was sharply reversed during the 1970s and 1980s, after the advent of stricter environmental regulations in the OECD countries. Concurrently, toxic intensity in LDC manufacturing grew quickly. Lucas et al (1992), Low and Yeats (1992) also confirm this displacement hypothesis.

Agras and Chapman (1999) and Suri and Chapman (1998) analyse the composition of international trade and observe that manufacturing goods exporting countries tend to have higher energy consumption. They find the poor and rich countries to be net exporters and net importers of pollution-intensive goods, respectively. Therefore, the inverted U-shaped *EKC* curve might partly be the result of changes in international specialization under which poor countries engage in *dirty* and energy intensive production while rich countries specialize in *clean* and service intensive production, without effectively any change in the consumption patterns.

It should be mentioned that a polluting activity in a high-income country normally faces higher regulatory costs[15] than its counterpart in a developing country (Mani and Wheeler (1998)). Under these circumstances the pollution intensive industries will have a natural tendency to migrate to countries with weaker environmental regulations (Copeland and

[14] Gawande et al., (2000, 2001) provide evidence that migration is a contributing factor behind the *EKC* especially for US hazardous waste (See also Berrens et al., 1997).

[15] This creates an incentive for at least some highly polluting industries to relocate. The firms are relocated to low-income countries with weak environmental regulation. Rising capital outflows force governments in high-income countries to begin relaxing environmental standards.

Taylor (1995)). This is referred to as the *Pollution Haven Hypothesis (PHH)* (See, Bommer (1999), Cole (2003, 2004)). The *PHH* refers to the possibility that multinational firms, particularly those engaged in highly polluting activities, relocate their polluting production activities to countries with lower environmental standard. In other words, the *PHH* basically suggests that countries having stricter environmental standard will lose all the *dirty industries* and poor countries (i.e., those having poorer environmental standard) will get them all. On the contrary, the *factor endowment hypothesis* (FEH) asserts that under free trade the differences in endowments (or technology) determine trade between two countries. Under this view capital-abundant countries tend to export capital-intensive goods, regardless of differences in environmental policy (Copeland and Taylor 2004). According to the FEH[16] polluting industries will concentrate in affluent countries, which also tend to be capital abundant. This is because polluting industries are typically also capital intensive and thus affluent capital-abundant countries having a comparative advantage in these industries. In this context, it should be noted that the differences in environmental policy and differences in factor endowments might jointly determine the comparative advantage in trade. Possibly, the FEH and PHH counteract and offset each other (Copeland and Taylor 2004). Antweiler et al., (2001) use the interaction of openness with relative income per capita and relative capital-labour ratio. Their estimated effect is quite small indicating that the FEH and PHH counteract and potentially tend to offset each other (see also Temurshoev 2006). The basic characteristics of a country and its dominating comparative advantage determine how trade liberalization influences its sectoral composition and consequently environmental outcomes (Copeland and Taylor 2004, Liang 2006).

Moreover, globalization, by increasing competition for investment, may trigger the environmental *race to bottom* (Wheeler (2000)). Poor economies may be able to improve their environmental quality as investment raise their income levels. Thus, globalization may facilitate pollution reduction. In fact, *the bottom* rises with economic growth. Tisdell (2001) points out that globalization can be a driving force for global economic growth. Yet opinion is divided about the benefits of this process. The global economy raises the issue of potential conflicts between two powerful current trends – viz., the worldwide acceptance of market oriented economic reform process on the one hand, and environmental protection on the other.

2.4.4. Market Mechanism

World Bank (1992) asserts that the existence of a self-regulatory market mechanism for traded natural resources may prevent environmental degradation. The argument underlying this assertion runs as follows. Early stages of growth are often associated with a heavy exploitation of natural resources due to the relative importance of the agricultural sector. This tends to reduce the stock of natural capital of an economy over time. Efficiency of use of natural resources increases only after a threshold stage of development has been crossed.

---

[16] Under free trade the capital abundant country exports the capital-intensive (dirty) goods, which stimulates its production, thus raising pollution in the capital abundant country. Conversely, pollution falls in the capital-scarce country as a result of contraction of the production of pollution-intensive goods, since there is no comparative advantage of producing polluting goods in the developing world (Temurshoev 2006, Liang 2006, Mukhopadhyay et al 2005).

Then markets for environmental resources develop and prices begin to reflect the value of natural resources. Consequently, the rising price of natural resources reduces their exploitation at later stages of growth. Moreover, higher prices of natural resources also contribute to accelerate the shift toward less resource-intensive technologies (Torras and Boyce (1998)). Hence, not only induced policy interventions, market signals can also explain the shape of the *EKC*.

Now *EKC* has become a standardized notion in technical conversations about environmental policy. Strong policies and institutions in the form of more secure property rights, better enforcement of contracts and effective environmental regulations can help flatten the *EKC* (Panayotou (1997), Ezzati et al., (2001), Magnani (2000), Gangadharan and Valenzuela (2001)). Most of the empirical evidences suggest that environmental problems may be solved at higher levels of income only for some specific environmental quality indicators.

2.5. A Brief Review of Theoretical Analysis

There are some conceptual arguments that make the *EKC* conceivable from a theoretical viewpoint. Recently some attempts have been made to explain the *EKC* hypothesis theoretically. There are basically two major strands within the theoretical *EKC* literature – one stresses shifts in the use of production technologies which differ in their production intensity (Stokey 1998) and the other focuses on the characteristics of the abatement technology (John and Pecchenino (1994), Selden and Song (1995)). Recently, Brock and Taylor (2004) amend the Solow growth model (which is known as the Green Solow model) to include emission, abatement and stock of pollution. Assuming a rate of technological progress in abatement, they show that an *EKC* may result along the transition to the balance growth path.

Another prominent contributions is the static Andreoni and Levinson (2001) and Lieb (2002) models. Lieb (2002) generalizes Stokey's (1998) model and argues that satiation in consumption is needed to generate *EKC*. Andreoni and Levinson (2001) show that economies of scale in abatement are sufficient to generate *EKC*. They derive it directly from technological link between consumption of a desired good and abatement of its undesirable byproduct. However, the abatement expenditure may not be a determining factor behind the *EKC* for long-lived pollutants like hazardous waste that are neither easily abated nor can be shifted elsewhere. In this context, Gawande et al., (2001) develop a theoretical model of *EKC* based on the perfect mobility of households. In addition, Lopez (1994) and Bulte and Soest (2001) develop models for the depletion of natural resources such as forest or fertility of agricultural land. These models generate *EKC*s under appropriate assumptions (See also Stern (2004)).

Other theoretical contributions to this literature include Chaudhuri and Pfaff (1998), who posit a particular mechanism, *bundled commodities*, to explain *EKC*, whereas, Lopez and Mitra (2000) try to explain the observed relationship between development and environmental quality in terms of corruption. They show that pollution level corresponding to corrupt behaviour is always above the socially optimal level and the *turning point* takes place above that corresponding to the social optimum.

2.6. Conclusion

The Environmental Kuznets Curve model has elicited conflicting reactions from researchers and policymakers. The stakes in the *EKC* debate are high for both developing and developed countries. Among a multiplicity of possible outcomes, an inverted-U pattern can only be obtained under specific circumstances. More importantly, this requires attention to the array of factors that form the economic-environmental system, rather than a single dominant one. So, *EKC*-analysis has significant deficiencies. Evidence for the existence of the *EKC* is quite inconclusive. However, the subject is open-ended and *EKC*-analysis will continue to be widely used.

Chapter 3

A THEORETICAL BASIS FOR DEVELOPMENT WITH ENVIRONMENT*

3.1. INTRODUCTION

Worldwide environmental degradation makes people worried about the issue of the link between economic growth and environmental degradation. A sizeable literature[17] on that subject, both theoretical as well as empirical, has grown in recent period. Among the vast empirical studies an important finding is the Environmental Kuznets Curve (*EKC*), viz., the inverted U-shaped relationship between pollution and economic growth. Environmental quality deteriorates initially and improves with economic development in later stage. The literature has mostly considered *EKC* as an empirical phenomenon. In order to strengthen connection between theoretical and empirical analyses, one needs models and stylized facts. One should analyze theoretical model(s) and observe under what conditions *EKC* is generated. An empirical regularity provides a relevant dimension for calibration of environmental aspects of growth model. This is useful to evaluate models on the basis of their analytical tractability and that of their compatibility with facts. The empirical evidence depends only on the reduced-form rather than the structural-forms. So, the question, why pollution-income follows this inverted U-shaped curve is not yet resolved fully.

Lopez (1994) and Selden and Song (1995) consider exogenous technological change and pollution is generated by production. The relationship between pollution and income level depends on the elasticities of substitution of goods and the risk preference of household (Lopez 1994). John and Pecchenino (1994) develop model based on overlapping generations where pollution is generated by consumption rather than production. Stokey (1998) allows endogenous technological change and Lieb (2002) generalizes Stokey's model and argues that satiation in consumption is needed to generate *EKC*. Andreoni and Levinson (2001) show that economies of scale in abatement are sufficient to generate *EKC*. They derive it directly from technological link between consumption of a desired good and abatement of its undesirable byproduct. The role of abatement expenditure is crucial to reduce the pollution in production side (Selden and Song

* Chapter is based on Dinda (2005).

[17] See, for example, Agras and Chapman (1999), Bimonte (2002), Cole et al., (1997), Dinda et al., (2000), Gawande et al., (2000), Grossman and Krueger (1995), Munasinghe (1999), Pasche (2002), Rothman (1998), Selden and Song (1994), Shafik (1994), Suri and Chapman (1998), Tisdell (2001), World Bank (1992).

(1994), however, the abatement activity starts when a considerable capital stock is achieved (Selden and Song (1995)). In addition, Lopez (1994) and Bulte and Soest (2001) develop models for the depletion of natural resources[18] such as forest or fertility of land. Thus, these models generate *EKC*s under appropriate assumptions. Recently, Stern (2004) reviews the latest theories and Brock and Taylor (2004) discuss that pollution declines in high-income countries due to technological change.

This paper lays out an explanation for the *EKC*, not only that it also discusses the dynamics of the *EKC* in the framework of endogenous growth model[19]. A distinctive feature of the model is that the environmental capital enters into the utility function as well as the production function. Thus, this paper is slightly different from the existing literature[20]. Most of the existing papers have focused on the amenity value of environment without considering the environment as a productive asset. One important and unique feature of the production function is the substitutability of man-made capital (physical and human capital) and natural capital (i.e., environmental assets) that ensures long run growth. It is closely connected to the view of Tahvonen and Kuuluvainen (1993) – 'a model of economic growth with stock pollution allows smooth substitution between emissions and capital'. Pollution is endogenous in our model and abatement activity plays a crucial role to check the environmental degradation. However, our model actually combines stock of capital, pollution, stock of environment (natural capital) in an endogenous growth model. In addition, it discusses the transitional dynamics rather than just steady state.

The rest of the paper is organized as follows: an endogenous growth model is built up on the basis of capital and environmental stock in section 2, for analytic tractability specific models are used in section 3, the steady state condition is discussed in section 4, the existence of *EKC* is examined in section 5 and finally, the conclusion is drawn.

3.2. Model Setup

Consider a one-good (closed) economy for which environment, E, understood as a stock variable, affects production level and utility of the representative agent. For simplicity, we consider this economy consists of a single economic agent[21] who acts as producer as well as consumer (or we can think of a central planner's problem). The representative agent maximizes her present value of utility (or welfare of the society):

$$\text{Maximize}\, W = \int_0^\infty e^{-\rho t} U(C(t), E(t))dt\,;\; U_C, U_E > 0;\; U_{CC}, U_{EE} < 0;\; U_{CE} > 0. \qquad (1.1)$$

[18] See, Krautkraemer (1985), Tahvonen and kuuluvainen (1993).

[19] See, Aghion and howitt (1998), Mulligan and Sala-i-Martin (1993), Barro and Sala-i-Martin (1995), Bovenberg and Smulders (1995), Elbasha and Roe (1996), Michel and Rotillon (1995), Beltratti (1996) etc.

[20] There is an extensive literature on pollution and growth, including early papers by Keeler et al., (1972), D'Arge and Kogiku (1973) Forster (1973), Gruver (1976) and more recent work by Tahvonen and kuuluvainen (1993), Gradus and Smulders (1993), John and Pecchenino (1994) etc.

[21] We abstract from all issues concerning population growth and labour supply by assuming a constant labour supply normalized to equal unity. Thus, all the variables should be interpreted as variables per capita.

Where C, E and ρ(>0) are composite consumption bundles, stock of environment (representing the stock of natural resources: land, air, water, flora and fauna, etc.) and rate of time preference, respectively.

Production function of the economy[22] is

$$Y = f(K_y, E);\ f_k,\ f_E > 0;\ f_{kk}, f_{EE} < 0. \tag{1.2}$$

Where K denotes composite capital stocks (physical and human capital) and E is environmental stocks. In production function, E also captures the productivity effect of the environmental quality (for example, the effect of air and water quality on health[23], etc.). E is not the choice variable to individual agents. Representative agent views the environmental stock (E) which is given, although E(t) varies in the aggregate. The two inputs, K and E, are essential for production.

Let γ (>0) be the rate of pollution (i.e., emission or degradation of environment per unit of output produced). Pollution (or emission) is fixed[24] proportion of output or production.

$$P = \gamma Y,\ \gamma > 0. \tag{1.3}$$

In general, the quality of the environment is itself endogenous. In aggregate, E is a stock that is depleted gradually by the flow of pollutants (or emission) P, as

$$\dot{E} = -P \text{, or } \dot{E} = -\gamma f(K_y, E) \tag{1.4}$$

This implies that stock E is decreasing over time. So, as $t \to \infty$, $E \to 0$; therefore, as $f(k_y, 0) = 0$, the economy will collapse. It is true that significantly high enough pollution damages production. For example, acid rain destroys natural capital like forests, soil, river water etc and also damages (indirectly) human capital. Thus, pollution affects output by damaging the inputs used to produce the output. Here lies the importance of abatement activity which upgrades (or clean up) environment. For this abatement activity, economy needs some capital[25] (K_E, say) which helps to improve environmental quality. We assume that abatement activity depends on man-made (physical & human) capital alone, i.e.,

$$A = h(K_E) \tag{1.5}$$

[22] In Forster (1973) model, the production function depends only on the stock of capital in the economy, in addition to this, our model adds natural resource or environment as input.

[23] Human health may be harmed due to pollution and thus, economic productivity may be lost. See also Gangadharan and Valenzuela 2001.

[24] γ may be a decreasing function of technology (or R & D). For simplicity, we assume a constant γ.

[25] It should be mentioned that this model is a model for a country developing technology endogenously without the possibility of adopting a ready made foreign technology. See, Parente and Prescott (2000) for foreign technology, especially for technology leader.

So, net change in E over time is: $\dot{E} = A - P$. Now, at each point of time, representative agent allocates optimally her stock of capital for both commodity production and abatement activity. Given total amount of capital ($K = K_y + K_E$), optimal sectoral allocation of capital for both sectors at any moment of time is governed by $\frac{\partial Y}{\partial K_Y} = q \frac{\partial A}{\partial K_E}$, where q is the price of environment relative to physical capital. This is true only in pollution free world, whereas in the polluted world, $(1 - q\gamma)\frac{\partial Y}{\partial K_Y} = q \frac{\partial A}{\partial K_E}$ is true, where $q\gamma \frac{\partial Y}{\partial K}$ is the cost of pollution (or externality). This condition states that the net marginal product of capital should be same in two sectors.

Now, $K = K_y + K_E$, or $K_y = \theta K$, where $\theta = \frac{K_y}{K}$. Therefore, $K_E = (1-\theta)K$. (1.6)

Let *θ(t) (0<θ(t)<1)* portion of capital stock be used for commodity production at time *t* and the remaining *(1-θ(t))* portion be used for upgrading (or cleaning) environment[26]. Now, Production function and environmental upgrading function can be re-expressed as $Y = f(\theta K \ , E)$ and $A = h((1-\theta)K \ )$, respectively. So, the accumulation constraints are (for simplicity, we assume, natural depreciation rates are zero for both the stocks, i.e., $\delta_i = 0$, $i = K, E$).

$$\dot{k}(t) = f(\theta(t)K(t), E(t)) - C(t) \tag{1.7}$$

and $\dot{E}(t) = h((1-\theta(t))K(t)) - \gamma f(\theta(t)K(t), E(t))$ (1.8)

Clearly, the first constraint (eq. 1.7) relates to physical capital accumulation while the second (eq. 1.8) relates to net environmental change due to production and environmental upgrading. The infinite time horizon inter-temporal consumption choice problem for this economy may be specified as

$$\text{Maximize } W = \int_0^\infty e^{-\rho t} U(C(t), E(t))dt$$

subject to the accumulation constraints (1.7) & (1.8). In this economy, representative agent has control over her consumption C as well as capital allocation θ for both commodity production and upgrading environment. Treating *C(t)* and *θ (t)* as control variables and *K(t)* and *E(t)* as state

[26]Our model is closely related to Gruver (1976). Using neoclassical growth model, Gruver (1976) examines the optimal division of investment between pollution control capital and directly productive capital.

variables (and suppressing the time suffixes of the variables), the current value Hamiltonian for the above optimization problem may be written as

$$H = U(C,E) + \lambda[f(\theta K,E) - C] + \mu[h((1-\theta)K) - \gamma f(\theta K,E)] \tag{2}$$

Where $\lambda(t)$ and $\mu(t)$ are two co-state variables (i.e., the shadow prices) corresponding to state variables *K* and *E*, respectively. The necessary conditions for an optimal solution of the problem are given in appendix A. Now we rewrite first two conditions as:

$$\frac{\delta H}{\delta C} = U_C - \lambda = 0 \Rightarrow U_C = \lambda \tag{2.1}$$

$$\frac{\delta H}{\delta \theta} = \lambda f_K - \mu(h_K + \gamma f_K) = 0 \Rightarrow \mu = \frac{\lambda f_K}{h_K + \gamma f_K} \tag{2.2}$$

Where U_C, f_K and h_K denote the marginal utility of consumption, marginal productivity of capital in commodity production and in environment upgrading, respectively. While condition (2.1) asserts the consumption-accumulation indifference at the margin, condition (2.2) requires the value of marginal product of capital in commodity production is equal to the value marginal of environmental quality loss due to withdrawal of capital from environment upgrading. Rearranging the conditions given in appendix A, we obtain consumption path[27]

$$\frac{\dot{C}}{C} = \frac{-U_C}{CU_{CC}}\left[\frac{f_k h_k}{h_k + \gamma f_k} - \rho + \frac{EU_{CE}}{U_C}\frac{\dot{E}}{E}\right] \tag{3a}$$

Where U_{CC}, U_{CE} being the second ordered partial derivatives of $U(.)$. Note that the above condition suggests those optimal time paths of *C* and *E* should generally be interdependent. If, however, $U_{CE}(U_{CC})$ turns out to be identically zero, the optimal time path of *C* (*E*) will be autonomous and the nature of the optimal time path of *E* (*C*) will depend upon what the optimal path of the other variable is.

[27] Since, the utility function and production function are strictly concave, it follows from Cass (1965) that eq. (A2.1) through eq. (A2.6), along with the transversality conditions $\lambda(t)e^{-\rho t} \rightarrow 0$ as $t \rightarrow \infty$, and $\mu(t)e^{-\rho t} \rightarrow 0$ as $t \rightarrow \infty$, are a sufficient characterization of the unique optimum path (solving the planner's problem).

3.3. Model Specification

Now, we consider specific functions rather than general function, with a parametric specification of the functions for analytic tractability. It helps to identify some key concepts.

The utility function is given by customary form of constant elasticity of marginal utility with respect to its argument. The utility function[28]:

$$U(C,E) = \frac{(C^{1-\upsilon}E^{\upsilon})^{1-\sigma} - 1}{1-\sigma} \tag{3.1}$$

The parameter υ ($0 < \upsilon < 1$) and (1-υ) represent preferences towards the environment and consumption, respectively. Here, σ is the usual elasticity of the marginal utility with respect to its argument and it is also known as intertemporal elasticity of substitution. Imposing condition $U_{CE} > 0$, implies that $\sigma < 1$. Conditions $U_{EE} < 0$, $U_{CC} < 0$, imply that $\upsilon(1-\sigma) < 1$ and $\sigma > 0$.

The technology for upgrading environmental quality or abatement function is assumed to be linear; in other words, constant returns to capital are assumed. Abatement function is

$$A = A_1 K_E = A_1(1-\theta)K \tag{3.2}$$

This is an AK type technology for upgrading the environmental quality or abatement activities[29]. With these specifications, the equilibrium growth rate of consumption (equation 3a) will be

$$\frac{\dot{C}}{C} = \frac{1}{\sigma}\left[\frac{A_1 f_K}{A_1 + \gamma f_K} + \upsilon(1-\sigma)\left(\frac{\dot{E}}{E}\right) - \rho\right] \tag{3b}$$

This is the optimal path of consumption of the economy. In this model, we find that economic growth ($\frac{\dot{C}}{C}$) is affected by the net change in environmental quality (or $\dot{E}$) or environmental growth rate (i.e., $\frac{\dot{E}}{E}$). Since, $\upsilon > 0$ and $\sigma < 1$ imply that $\upsilon(1-\sigma) > 0$ which is the common assumption in the literature.

Remark 1: *An improvement in Environmental quality increases the marginal utility of consumption, raising the incentive to consume at all times and thus to accumulate.*

---

[28] This is different from others, for example, Asako (1980) analyzes the interactions between capital accumulation and pollution under the constant utility, which depends on per capita consumption and pollution. See, also Becker (1982).

[29] It differs from J type function. Selden and Song (1995) provide a theoretical basis for recent empirical research on J curves for abatement effort and inverted U curves for pollution. The result of Andreoni and Levinson (2001) crucially depends on increasing returns in abatement function. See also Withagen (1995).

An improvement in environmental quality ($\dot{E} > 0$) raises the equilibrium growth rates, *ceteris paribus*. This is the situation in which both equilibrium growths increase and environmental quality improves. If utility function is an additive form (i.e., $U_{CE} = 0$), or if we drop E from utility function (eq.(1.1)), then eq.(3a) is the standard economic growth rate (i.e., $\dot{E}/E$ vanishes from eq. (3a)). Now if $\dot{E}/E < 0$, following eq.(3a), the optimum economic growth rate falls. Hence, the proposition follows

Proposition 1: *Economic growth remains at optimum provided the economy ensures the non-negativity of the environmental growth.*

The equilibrium growth rate of capital allocation (θ) is

$$\frac{\dot{\theta}}{\theta} = \frac{-1}{(\theta K) f_{KK}} \left[\frac{U_E}{U_C} \frac{(A_1 + \gamma f_K)^2}{A_1} + (A_1 + \gamma f_K) f_E + f_{KE}\, \dot{E} + f_{KK} (\theta \dot{K}) - f_{KK}^2 \right] \quad (3.3)$$

For analytical purpose, let us consider Cobb-Douglas type production function,

i.e., $Y = f(\theta K, E) = B(\theta K)^{\alpha} E^{1-\alpha}$ (3.4)

3.4. Steady State

This model has two control variables (C and θ) and two state variables (K and E). Hence, the phase diagram of the planning problem is four-dimensional. In general, the four-dimensional problem is hard to study analytically. It will be easy to analyze the problem if we reduce four-dimensional problem to three or two-dimension. Four variables can be reduced to three variables by normalizing or simply dividing by one variable, say E. After eliminating θ , we obtain c (C/E) and k (K/E), which are analyzed in two-dimensional plane (See, figure A.1 in Appendix B). For analytical purpose some restrictions are imposed, which may help to trace out the movements of E and K variables in E- K plane (See, Appendix C). The change of the control variables C and/or θ affect the locus of $\dot{E} = 0$ and $\dot{K} = 0$ in *E*- *K* plane (see, figure A.2 in Appendix B). There are several possible positions of $\dot{E} = 0$ and $\dot{K} = 0$, taking one specific situation, we concentrate and focus on the movement of the variables E and K in E-K plane such that one possible situation produces *EKC* (other cases are shown in figure A.3a and figure A.3b in Appendix B).

In this study, we show diagrammatically one particular initial value of K and E leads optimally towards steady state. With initial capital K_0 (very low) and environment E_0 (very high), one economy starts to grow and in the process of development, economy accumulates capital and reduces environmental stock over time (see, line 2, in Figure 1). In other words, there is a trade off between environment and capital, thus, economic growth is possible at the cost of environmental degradation. This is a natural phenomenon of an under developed economy which is ready to forgo environmental stock for economic growth. This developmental process continues till economy attains the steady state.

Let us consider it in different way, the capital, K, is scarce and environmental resource, E, is abandoned in under developed economy. The (shadow) price of E (i.e., μ) is unchanged but that of K (i.e., λ) increases. So, the relative price of environment, q, decreases. With the process of economic development in underdeveloped countries, K increases and E approximately remains constant or slowly declines, therefore, k (or, K/E) increases towards k^* (optimum capital stock) while q decreases. In the reverse situation, k declines with rising q. The economy attains a stable equilibrium at S (corresponding to k^*) following either line 2 (or 7) or line 4 in Figure 1. The usual standard proposition follows

Proposition 2: *If $k(0)<k^*$, k increases towards k^* as q decreases and if $k(0)>k^*$, k decreases towards k^* while q increases towards q^*.*

The steady state situation is $\frac{\dot{C}}{C}=\frac{\dot{K}}{K}=\frac{\dot{Y}}{Y}=\frac{\dot{E}}{E}=\frac{\dot{A}}{A}=\frac{\dot{\theta}}{\theta}=0$, where θ and C/K, C/E and K/E ratios are constant. In the transitional dynamics, the environment continuously falls (see, arrow line marked 2 and 7 in Figure 1) and halts at steady state where no more environmental deterioration takes place. So, inverted U-curve (or *EKC*) does not arise either at steady state or in transitional dynamics moving (along optimum path) towards steady state.

Remark 2: *In transitional dynamics, a trade-off between economic growth and environmental degradation should be optimal and that is the necessary condition for the convergence towards Steady State.*

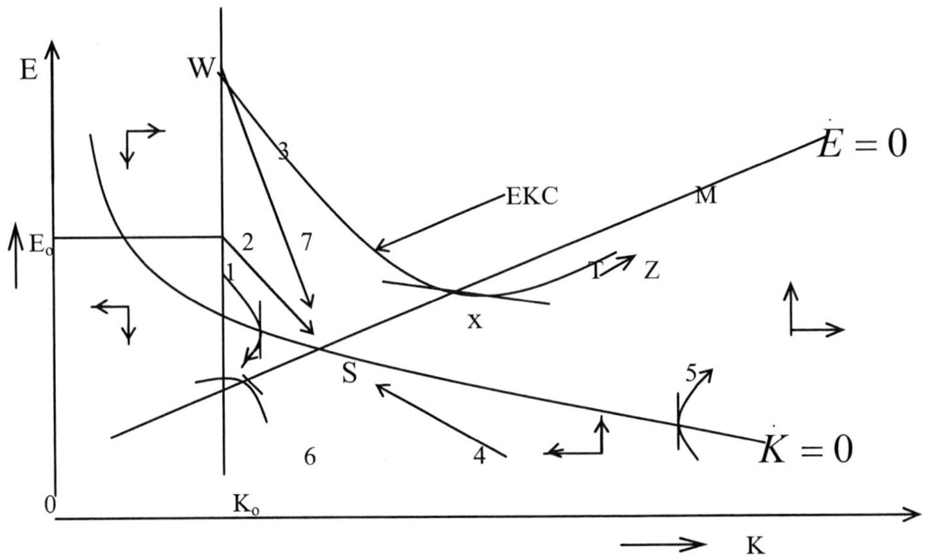

Figure 1. Transitional dynamics.

3.5. EXISTENCE OF EKC

The economic development at the cost of environmental degradation may continue for long time (along WS in Figure 1) or after some time, both E and K may fall (see curve 1) or rise (see, curve 3 in figure). Along line 2 (or 7) $\dot{K} > 0$ and $\dot{E} < 0$, with growth it moves towards $\dot{K} = \dot{E} = 0$ at S, whereas curve 3 has no tendency towards S or $\dot{K} = \dot{E} = 0$.

From the initial point W, the economy might follow optimal allocation of capital along WS (See, line 7 in Figure 1) which moves toward equilibrium point 'S'. So, the rate of environmental degradation is fixed along WS. Thus, the optimal path of E decreases monotonically as K increases and stabilizes at S. It should be mentioned that some empirical evidences confirm it, especially the municipal waste and CO_2 emission increase monotonically whereas the access of safe drinking water and sanitation improve steadily with rising level of income in most of the countries (World Bank (1992), Shafik (1994)).

It should be noted that only saddle path leads to the steady state and all other paths either violate the transversality conditions or hit constraints. In this context, we concentrate on curve 3 (in Figure 1), which follows the law of motion but can't be equilibrium for the reasons just mentioned. Economy will collapse (see, curve 1 in Figure 1) if the rate of environmental degradation is more compare to the optimal rate and exhibits an *EKC* (along WXTZ) if it is less. In Figure 1, the curve 3 depicts the *EKC* path in the off-steady state[30] and it is totally unstable because both the stocks (E and K) increase over time at Z point. At point X, the net change in environmental stock is zero (i.e., $\dot{E} = 0$) but stock of capital still increases (i.e., $\dot{K} > 0$) which is essential for continuing economic growth as well as to maintain $\dot{E} = 0$. Thus, the environment is stabilized[31] with capital accumulation.

Remark 3: *A sustainable development of an economy may continue only when* $\dot{K} > 0$ *and net environmental change remains unaffected (i.e.,* $\dot{E} = 0$*).*

Economic growth may be compatible with environmental protection, but this is not automatic. With an insufficient amount of resources allocation for environmental improvement, economic growth will tend to exhaust the environmental resource in long run. If society does not devote appropriate amount of resources to environmental improvement or protection, higher rate of growth[32] can only be temporary, in long run it will eventually be zero.

Suppose, WX path crosses the $\dot{E} = 0$ path at X then it moves along XTZ path. At point T or Z, both stocks grow (i.e., $\dot{E} > 0$) and $\dot{K} > 0$), so environmental quality will improve along XTZ

---

[30] The curve 3 follows law of motion but hits some constraints before reaching the steady state. Hence it is clear that the curve 3 exists only in non-optimal/sub-optimal situation. There exists infinite number of such curves but planner chooses that path which gives him/her maximum welfare.

[31] In this context it should be mentioned that Stephens (1976) allows exponential growth of per capita income with constant environmental quality.

[32] The optimal growth rate (income is defined in narrow sense) will be lower if pollution is more because increased abatement activities crowd out investment. In broad definition of GDP, which incorporated the value of abatement activities, long run growth rate will be constant and also leads to a sustainable development in the economy.

path[33]. Thus, in Figure 1 WXTZ path exhibits U shaped relationship between E and K in E - K plane. Initially environmental quality declines as the whole capital used up only in production process, nothing left for environmental protection. It is true that none the economy could thought about the environmental degradation (or constrained) in their planning at the early stages of economic development. After sometimes (at least after industrialization) economy should realize the consequence of it, and economy gradually increases the allocation of capital to protect the environment and possible to improve environment at a higher level of capital. Once the abatement expenditure is positive, according to Lieb (2002), it is sufficient to find *EKC*. It is clear that allocation of capital is insufficient for abatement activity along WX but sufficient or more along XTZ. Hence, a move from insufficient to sufficient allocation of capital for abatement activity is the basis for *EKC*. It should be mentioned that only one subset of sub-optimal situations generate *EKC*.

In this study, the policy variable θ (allocation of capital) plays the vital role for controlling environmental quality. In the process of economic development, allocation of capital for abatement activity monotonically raises over time and halts at optimum or steady state. Thus, generally the abatement activity becomes strong enough to restore the environmental quality. It will be strong only when a sufficient investment takes place in the abatement sector, then environmental quality improves. This is the basis for U-shaped relationship between environmental quality and economic growth.

Remark 4: *The abatement activity is the main force to stop the environmental degradation. So, optimal allocation of capital for abatement activity is the sufficient investment for curve down the pollution level.*

A sufficient resource has to be devoted for abatement activity and at some time it will become optimal to stop environmental degradation with capital accumulation, and economic growth may continues protecting environmental quality. It should be noted that the allocation of capital protects environment, which is crucial for the long run growth.

3.6. CONCLUSIONS

This paper provides a possible theoretical explanation for the Environmental Kuznets Curve in the framework of endogenous growth model. The *EKC* exists dynamically in the off-steady state. Generally, less developed economies use their whole stock of capital for commodity production that generates pollution, which damages their existing environmental stocks. This model suggests that each economy should allocate one part of their capital for abatement activity. Environmental degradation continues at early stage because of insufficient investment for abatement activity, but in later stage, sufficient investment (viz., optimal allocation of capital) prevents further degradation of environmental quality. Thus, to restore environmental quality, a sufficient investment is needed that is possible only when economy accumulates enough capital stocks. The shift from insufficient to sufficient investment for upgrading environment is the basis

[33] E increases maximum up to its initial level or "pristine level", then $\dot{E} = 0)$ but still $\dot{K} > 0$, that means economy grows. Further economic growth/development may lead to fall in E, i.e., $\dot{E} < 0)$. Truly, there is cycle for E, which may provide multiple equilibrium (See, Figure A.3a and Figure A.3b in Appendix B). *EKC* is observed in one phase of a cycle. Extending to another phase, we may get 'N' shape curve.

for curve down the pollution level, and thus, correspondingly forms the inverted U-shaped relationship between pollution and economic growth. Optimal allocation of capital for abatement activity is the sufficient investment demand to curve down the pollution level.

APPENDIX A

Followings are the necessary conditions for an optimal solution of the problem:

$$\frac{\delta H}{\delta C} = U_C - \lambda = 0 \Rightarrow U_C = \lambda \tag{A2.1}$$

$$\frac{\delta H}{\delta \theta} = \lambda f_K - \mu(h_K + \gamma f_K) = 0 \Rightarrow \mu = \frac{\lambda f_K}{h_K + \gamma f_K} \tag{A2.2}$$

$$\dot{K} = \frac{\delta H}{\delta \lambda} \Rightarrow \dot{K} = f(\theta K, E) - C \tag{A2.3}$$

$$\dot{\lambda} = -\frac{\delta H}{\delta K} + \rho\lambda = (\mu\gamma - \lambda)\theta f_K - \mu(1-\theta)h_K + \rho\lambda \tag{A2.4}$$

$$\dot{E} = \frac{\delta H}{\delta \mu} \Rightarrow \dot{E} = h((1-\theta)K) - \gamma f(\theta K, E) \tag{A2.5}$$

$$\dot{\mu} = -\frac{\delta H}{\delta E} + \rho\mu = -U_E + (\mu\gamma - \lambda) f_E + \rho\mu \tag{A2.6}$$

Where U_C, f_K and h_K denote the marginal utility of consumption, marginal productivity of capital in commodity production and in environment upgrading, respectively. While condition (A2.1) asserts the consumption-accumulation indifference at the margin, condition (A2.2) requires the value of marginal product of capital in commodity production is equal to the value marginal of environmental quality loss due to withdrawal of capital from environment upgrading. Equation (A2.3) and (A2.5) are the motion equations of the state variables (K and E). Equations (A2.4) and (A2.6) are the well-known arbitrage conditions for capital formation and environment upgrading, which ensure that the economy would be indifferent between consumption and saving, and between production and environment improvement at the margin, respectively.

APPENDIX B

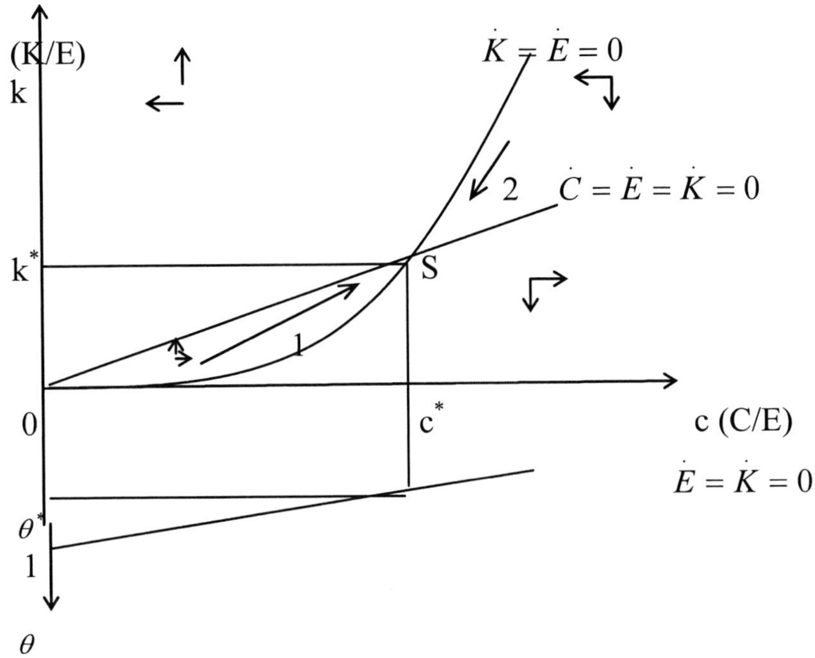

Figure A.1. Steady State and Optimal values.

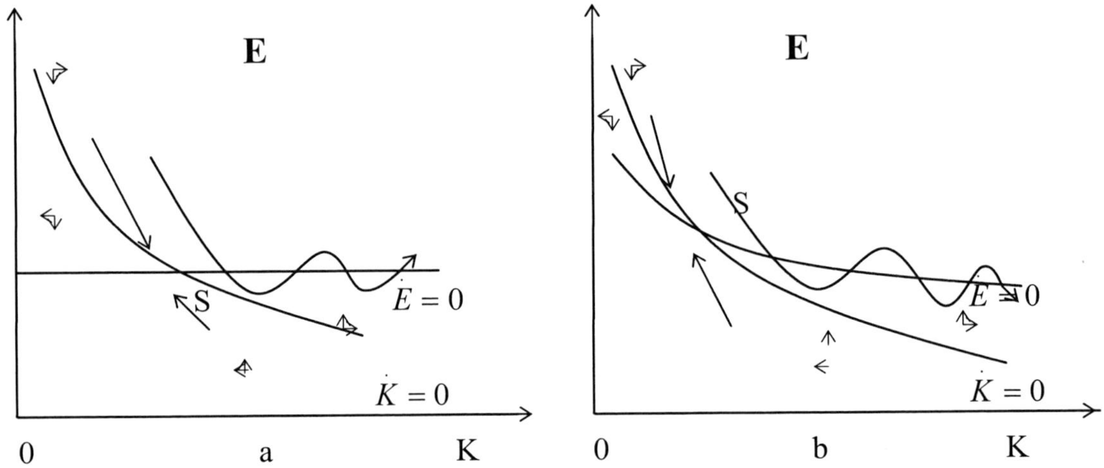

Figure A.3. Transitional dynamics and multiple equilibrium.

APPENDIX C

Optimum Values at Steady State

Now we define c and k as $c = \dfrac{C}{E}$ and $k = \dfrac{K}{E}$. Solving $\dot{K} = \dot{E} = 0$ and $\dot{C} = \dot{K} = \dot{E} = 0$, we get $k = \dfrac{1}{B^{\frac{1}{\alpha}}} c^{\frac{1}{\alpha}} + \dfrac{\gamma}{A_1} c$ and $k = \alpha\left(\dfrac{1}{\alpha} - \dfrac{\gamma}{A_1}\right)c$, respectively. At steady state, the optimal values are $c^* = B^{\frac{1}{1-\alpha}}\left(\dfrac{\alpha}{\rho} - (1+\alpha)\dfrac{\gamma}{A_1}\right)^{\frac{\alpha}{1-\alpha}}$ and $k^* = \alpha\left(\dfrac{1}{\alpha} - \dfrac{\gamma}{A_1}\right)c^* = \alpha B^{\frac{1}{1-\alpha}}\left(\dfrac{1}{\rho} - \dfrac{\gamma}{A_1}\right)\left[\dfrac{\alpha}{\rho} - (1+\alpha)\dfrac{\gamma}{A_1}\right]^{\frac{\alpha}{1-\alpha}}$, and the optimal value of θ is $\theta^* = 1 - \dfrac{\gamma}{A_1}\left(\dfrac{C}{K}\right)^*$ or, $\theta^* = 1 - \dfrac{\gamma}{A_1} c^* = 1 - \dfrac{\gamma}{A_1}\left[B^{\frac{1}{1-\alpha}}\left(\dfrac{\alpha}{\rho} - (1+\alpha)\dfrac{\gamma}{A_1}\right)^{\frac{\alpha}{1-\alpha}}\right]$.

These optimal values (θ^*, c^* and k^*) are also derived graphically (See Figure A.1 in Appendix B). It should be noted that the optimal allocation of capital (θ^*) depends on the pollution rate (γ), available technologies in production (B) and abatement activity (A_1), share of capital in production (α), and the discount rate (ρ).

Existence of Saddle Path

It is easy to analyze the existence of a saddle path in *k-c* plane. Solving $\dot{C} = \dot{K} = \dot{E} = 0$ in *k-c* plane (See line 1 and 2 in figure A.1), we obtain equation of the saddle path i.e., $A_1 k = m + \gamma c$, where $m = \left(\dfrac{A_1^{\alpha}}{(1-\gamma)\alpha B}\right)^{\frac{1}{\alpha-1}}$. These k and c ratios are linearly related, which is shown in figure A.1 in appendix B.

Now, we analyze the existence and stability of saddle path of four-dimensional problems in two-dimensional *E-K* plane. It is hard to analyze four variables in two-dimensional plane. Some restrictions are needed to analyze it.

$$\dot{K} = f(\theta K, E) - C$$

$$\phi_K \equiv \frac{d\dot{K}}{dK} = f_K + f_\theta \frac{d\theta}{dK} + f_E \frac{dE}{dK} - \frac{dC}{dK}$$

$$\phi_\theta \equiv \frac{d\dot{K}}{d\theta} = f_K \frac{dK}{d\theta} + f_\theta + f_E \frac{dE}{d\theta} - \frac{dC}{d\theta}$$

$$\phi_E \equiv \frac{d\dot{K}}{dE} = f_K \frac{dK}{dE} + f_\theta \frac{d\theta}{dE} + f_E - \frac{dC}{dE}$$

$$\phi_C \equiv \frac{d\dot{K}}{dC} = f_K \frac{dK}{dC} + f_\theta \frac{d\theta}{dC} + f_E \frac{dE}{dC} - 1$$

$$\dot{E} = A_1(1-\theta)K - \gamma f(\theta K, E)$$

$$\psi_K \equiv \frac{d\dot{E}}{dK} = A_1(1-\theta) - \gamma f_K - (A_1 K + \gamma f_\theta)\frac{d\theta}{dK} - \gamma f_E \frac{dE}{dK}$$

$$\psi_\theta \equiv \frac{d\dot{E}}{d\theta} = (A_1(1-\theta) - \gamma f_K)\frac{dK}{d\theta} - (A_1 K + \gamma f_\theta) - \gamma f_E \frac{dE}{d\theta}$$

$$\psi_E \equiv \frac{d\dot{E}}{dE} = \{A_1(1-\theta) - \gamma f_K\}\frac{dK}{dE} - \{A_1 K + \gamma f_\theta\}\frac{d\theta}{dE} - \gamma f_E$$

$$\psi_C \equiv \frac{d\dot{E}}{dC} = 0$$

Now,

$\phi_K > 0$ if $f_K > f_\theta \frac{d\theta}{dC} + f_E \frac{dE}{dK} - \frac{dC}{dK}$, $\frac{d\theta}{dK}|_{\dot{K}=0} < 0$, $\frac{d\theta}{dK}|_{\dot{E}=0} < 0$,

$\phi_E > 0$ if $f_E + \frac{dC}{dK} > f_K \frac{dK}{dE} + f_\theta \frac{d\theta}{dE}$, $\frac{dC}{dK}|_{\dot{K}=0} > 0$, at low K, and <0, at high K.

$\psi_K > 0$, $\psi_E \leq 0$

Under these restrictions, in (E, K) plane, the determinant is negative, i.e.,

$$D = \begin{vmatrix} \phi_K & \psi_K \\ \phi_E & \psi_E \end{vmatrix} < 0.$$

This suggests that there exist a saddle path in *E-K* plane under above-mentioned restrictions. It should be noted that if once the economy attains steady state $(\frac{\dot{C}}{C} = \frac{\dot{K}}{K} = \frac{\dot{E}}{E} = \frac{\dot{\theta}}{\theta} = 0)$, economy maintains it.

Chapter 4

AIR QUALITY AND ECONOMIC GROWTH: AN EMPIRICAL STUDY*

4.1. INTRODUCTION

Worldwide deterioration of environmental quality made many feel concerned about the issue and a sizeable literature on the pollution-income growth relationship has grown in the recent period. The World Development Report-1992 presents cross sectional evidences on the relationship between different indicators of environmental quality and per capita national income across countries. Other studies (e.g., Selden and Song (1994), Shafik (1994), Grossman and Krueger (1995), McConnell (1997), Carson et al., (1997), Suri and Chapman (1998), and Rothman (1998)) have found inverted U-shaped relationship between environmental degradation and income. The common point of all these papers is the assertion that the environmental quality deteriorates initially and then improves as an economy develops. This inverted U-shaped relationship between environmental deterioration and economic growth has been called the Environmental Kuznets curve (EKC). Explanation of the EKC, as already discussed in Chapter 2, has been pursued on many lines. Two major explanations are as follows: (i) use of environment as a major source of inputs and a pool for waste assimilation increases at the initial stage of economic growth, but as a country grows richer, structural changes take place which results in greater environment protection; and (ii) viewed as a consumption good, the status of environmental quality changes from a luxury to a necessary good as an economy develops. Phenomena like structural economic change and transition, technological improvements and rise in public spending on environmental R & D with rising per capita income level are considered to be important in determining the nature of relationship between economic growth and environmental quality. Grossman and Krueger (1995), using cross-country city level data on environmental quality, found support for the EKC hypothesis with peaks at a relatively early stage of development[34]. However, no such peak was observed for the heavier particles. Shafik (1994) also estimated the *turning point* for suspended particulate matter (SPM) to be at per capita GDP $ 3,280. Selden and Song (1994)

---

* This Chapter is based on Dinda et al., (2000).

[34] Namely, for lighter particles (i.e., smoke) and sulphur dioxide (SO_2) the observed peak corresponded to per capita GDP level of US$6,151 and 4,053, respectively. It may be noted that the per capita GDP values reported here and elsewhere in this chapter are measured at 1985 US prices.

used aggregate emission data (rather than the data on concentration of pollutant in the atmosphere, as used in many studies including the present one) and estimated peaks for air pollutants at per capita GDP levels greater than $ 8,000. The results of Cole et al., (1997) tend to suggest that meaningful EKC's exist only for local air pollutants. Vincent (1997) analysed the relationship between pollution and income level using time series data for Malaysia. His results, which contradict the findings obtained from the cross-country panel data, were thought to reflect the consequences of non-environmental policy decision. Carson et al., (1997) also obtained inverse relationship between per capita income and emission for seven major types of air pollutant in 50 US states. Further, they observed greater variability of per capita emission for the lower income states (which possibly suggests that the individual US states follow widely divergent development paths). Kaufmann et al., (1998) found a U-shaped relationship between income and atmospheric concentration of SO_2, and an inverted U-shaped relationship between spatial intensity of economic activity and atmospheric SO_2 concentration.

Socio-political conditions (Torras and Boyce (1998), Panayotou (1997)) are also found to have significant effects on environmental quality. Thus, while a faster economic growth may involve a higher environmental cost, a better institutional set up characterised by good governance, credible property rights, defined political rights, literacy, regulations etc. can create strong public awareness against environmental degradation and help protect the environment. Rothman (1998) and Suri and Chapman (1998) tried to explain the EKC phenomenon in terms of trade and consumption pattern differences of the developing and the developed countries. Their observation is as follows: Manufacturing industries (which are often more polluting) concentrate mostly in the less developed countries, whereas the less polluting high-tech industries (which are far less polluting) concentrate in the rich already industrialised countries due to the nature of the established pattern of international trade. Therefore, the rising portion of the EKC could be due to the concentration of manufacturing industrial activities in the developing countries and the declining portion of the EKC could be due to the concentration of less polluting high-tech industries in the developed world. Finally, household preferences and demand for environmental quality are also regarded as possible explanatory factors for the EKC phenomenon (McConnell (1997), Komen et al., (1997)). As the demand for environmental quality is income elastic, a strong private and social demand for a high quality environment in the developed countries would induce considerable private and public expenditures on environmental protection. Thus, whereas the rising portion of the EKC may be a manifestation of the substitution relationship between the demands for material consumption and environmental quality, the declining portion of the EKC may result as the substitution relationship turns to one of complementarity between the two kinds of demand.

The present chapter re-examines the EKC hypothesis of an inverted U-shaped relationship between environmental degradation and economic growth using the World Bank cross-country panel data on environment and per capita real GDP for the period 1979-90. Two different measures of environmental quality - viz., SPM and SO_2 have been used here[35]. An inverse relationship between the levels of air pollutant and per capita real GDP is observed. In case of SPM, a significant U-turn at a reasonably high per capita income level is found. This may be due to the fact that as income rises, the countries become more energy

[35] See World Development Report-1992, Table A.5, page 199.

intensive[36]. Recognising the possibility that the environmental quality of a country may, in addition to real per capita GDP, depend on the production technology, here we have attempted to examine if, in addition to the per capita GDP level, the production technique (i.e., capital-labour ratio) and the sectoral composition of GDP have any effect on pollution level[37].

In this context, it may be mentioned that whereas there has been numerous studies examining empirical validity of the EKC hypothesis, theoretical studies exploring the possible shapes of the income- pollution (environmental degradation) relationship are very few. Recently, Brock and Taylor (2004a) have successfully related the empirical regularity found in the EKC literature to the macroeconomic growth theory in the Solow model framework. They have extended Solow's model of economic growth by bringing in environment and its possible change due to production activity (which has been called the Green Solow Model) and investigated the relationship between economic growth and environmental outcomes. They provide a nice theoretical explanation of some puzzling features of the observed pollution and per capita income data. Their model produces several testable restrictions for the income-pollution relationship. For example, under certain assumptions (about population growth rate, (fixed) saving rate, (fixed) intensity of abatement, etc.) the Green Solow model produces a path for income per capita and environmental quality (flow of emission and stock of pollution) that trace out an EKC. More importantly, they show that such an EKC profile need not be unique and suggest alternative plausible scenarios. Applying the technique of convergence analysis (used in the empirical macroeconomic growth literature), they also derive a simple estimating equation that suggest convergence of per capita emissions across countries. In fact, in the Green Solow model the EKC is a necessary by-product of convergence to a sustainable growth path and the resulting EKC may be humped shaped or strictly declining which may be approximated by a quadratic income-pollution functional form in an empirical exercise, as has been used in the study presented in this Chapter.

This Chapter is organised as follows: Section 4.2 briefly explains the nature of data and the regression set up used in the present study and the regression results are presented and discussed in Section 4.3. Finally, Section 4.4 concludes the Chapter.

4.2. Description of the Data

The basic air pollution data on SPM and SO_2 used in the present study are obtained from World Development Report–1992. This report gives city-wise annual data on mean atmospheric concentration (microgram per cubic meter) of SPM and SO_2 separately for three time periods (viz., 1979-82, 1983-86 and 1987-90) for 33 countries classified into low, middle and high income groups. For each city in the sample, the data relate to the level of

[36] Using World Bank data (1986) on energy consumption per capita (kg. of oil equivalent) for high and low income countries the average propensity to consume for energy is estimated to be 0.446 kg./ $ for high income countries (viz., USA, UK, Canada, Finland, Norway, Netherlands, Japan, Germany, France) and 0.367 kg./ $ for low income countries (viz., Indonesia, Malaysia, Thailand, Philippines, Panama, Paraguay, Uruguay, Brazil, Morocco).

[37] The major determinants of environmental quality are specified to be resource endowment, income and technology. See Shafik (1994).

pollution either at the city centre or at the neighbourhood suburb. Further, the sites from where data were recorded in a city centre/suburb were classified as residential, commercial or industrial, as the case might be. The countries covered in the low income group were China, Egypt, Ghana, India, Indonesia and Pakistan; those covered in the middle income group include Brazil, Chile, Greece, Iran, Malaysia, Philippines, Poland, Portugal, Thailand, Venezuela and Yugoslavia, and finally, the high income group includes Australia, Belgium, Canada, Denmark, Finland, Germany, Hong Kong, Ireland, Israel, Italy, Japan, the Netherlands, New Zealand, Spain, the U.K. and the U.S.A. For the purpose of the present analysis, we have calculated country-wise annual mean concentration of SPM and SO_2 separately for residential and commercial centres for each of the three time periods mentioned above. The data thus constructed relate to 42 cities for SPM and 39 cities for SO_2 in 26 countries.

As regards the country-wise per capita income data, we have used the country-wise real per capita GDP (measured at a common set of international prices) available from the Penn World Tables (Summers and Heston (1994)). Since the pollution data are available city-wise for individual countries, ideally we should have some measure of city-wise per capita income. However, such income data being unavailable, we have used the real per capita GDP of the country (to which a specific city belongs) as a proxy for the per capita income of a city. Thus, for all the cities belonging to a country, the same country level per capita income has been used. As the city-wise pollution data are available separately for three time periods as already mentioned, we have used the average of yearly per capita incomes for a specific time-period as the measure of per capita income of that time period. Thus, the data set we have used in the present study is essentially of the nature of a panel data consisting of 42 cities in 26 countries and 3 time-periods[38]. Note that of the 26 countries represented in our data set, 15 belong to the high-income group. Thus, the present data set has a somewhat biased representation of countries with high income. The Table 4.1 presents a two-way summary of the distribution of countries and cities by per capita income level (PCGDP) and pollutant type.

In our empirical analysis reported in this chapter we have tried to explain the level of pollution in terms of production technique (as reflected by the capital-labour ratio for the economy as a whole) and sectoral composition of GDP of individual countries, in addition to PCGDP. Country-wise capital-labour ratios have been calculated on the basis of country-wise data on gross capital and employed labour force available in UN's National Accounts Statistics and ILO's Yearbook of Labour Statistics, respectively. Finally, country-wise data on sectoral composition of GDP have been obtained from the World Bank reports.

4.3. The Regression Set up and the Results

The primary focus of the present study is on the relationship between ambient air quality and real per capita GDP (*PCGDP*). Here we have considered a number of alternative quadratic functional forms of the basic regression model relating these two variables, viz.,

[38] To be precise, for SPM we have data for 42 cities in 26 countries, where as for SO_2 data for 39 cities in 26 countries.

Table 4.1. Distribution of sample by PCGDP level

Group**		Low PCGDP	Middle PCGDP	High PCGDP	All
SPM	No. of Countries	4	7	15	26
	No. of Cities	11	8	23	42
SO_2	No. of Countries	4	7	15	26
	No. of Cities	10	8	21	39

** As per World Bank guideline.

$$y_{it} = \beta_0 + \beta_1 x_{it} + \beta_2 x_{it}^2 \tag{4.1}$$

$$y_{it} = \beta_0 + \beta_1 \ln x_{it} + \beta_2 (\ln x_{it})^2 \tag{4.2}$$

$$\ln y_{it} = \beta_0 + \beta_1 x_{it} + \beta_2 x_{it}^2 \tag{4.3}$$

$$\ln y_{it} = \beta_0 + \beta_1 \ln x_{it} + \beta_2 (\ln x_{it})^2 \tag{4.4}$$

Table 4.2. Summary statistics of the suspended particulate matter for different groups and their combinations

Group	Variables	Mean	Variance	Correlation coeffi. between SPM & GDP	No. of Countries
c_1	SPM	127.97	12448	-0.91	14
	GDP	7034.6			
c_2	SPM	133.57	13572	-0.90	13
	GDP	7537.7			
c_3	SPM	141.15	15554	-0.87	7
	GDP	10226			
c	SPM	132.82	12692	-0.82	34
	GDP	7884.1			
r_1	SPM	186.17	16453	-0.87	6
	GDP	4960.8			
r_2	SPM	187.67	18966	-0.84	6
	GDP	5484.3			
r	SPM	194.99	17252	-0.79	14
	GDP	4912.1			
All	SPM	150.96	14499	-0.82	48
	GDP	7017.2			

Note: c_t is the group of countries with data from commercial areas of cities at time t, r_r is the same from residential areas. r_3 is not reported here because it contains only two countries with high level of SPM such that grand mean SPM exceed that of r_1 and r_2.

where y_{it} and x_{it} denote levels of air pollutant and real *PCGDP* for i^{th} country at t^{th} time period, respectively. These alternative quadratic functional forms of EKC may be regarded as approximation of the nonlinear functional forms of EKC that Brock and Taylor (2004a,b) have derived while studying the possibility of convergence to a sustainable growth path in their Green Solow model framework mentioned above and for all the four forms listed above, β_1>0 and β_2<0 will support the EKC hypothesis. In this context, it should be mentioned that empirically the parametrization of the above model may differ across countries and over time as the countries may have varying preferences, technology etc. and may be at different stages of development. The implicit identifying assumption is therefore that the countries have identical preferences, production and abatement technologies etc. and all of them move approximately along the same time path towards a steady state growth equilibrium, starting from very similar initial conditions and at a given point of time they are observed at different points on the path of transition.

Equations (4.1) – (4.4) for *SPM* and SO_2 have been estimated separately for residential and commercial locations and also using the combined data for the two types of locations for each of the time periods as well as for the pooled data for the three time periods. In each case, OLS estimation has been done assuming an additive spherical equation random disturbance term and the adequacy of the OLS estimation has been examined by performing some standard regression diagnostic tests like Jarque-Bera test of normality of residuals etc. In each case, for choosing the best-fitting form of equation (out of equations (4.1) – (4.4)), the adjusted coefficient of determination ($\overline{R}^2$) and the maximised log-likelihood value have been considered. In this context, it may be mentioned that since the data sets used here are essentially cross-sectional with observations corresponding to cities, commercial and residential locations, problem of spatial autocorrelation may be a possibility. As is known, presence of such spatial autocorrelation in the data is likely to affect the quality of the OLS estimate of the regression parameter vector. To be specific, if the equation disturbance term is spatially autoregressive, the OLS estimator of the parameter vector ceases to be efficient and the corresponding estimated sampling variance becomes biased. On the other hand, if the dependent variable is spatially autoregressive, the OLS estimator of the parameter vector associated with the explanatory variables may even be inconsistent (Anselin, 1988). In the case of the present exercise, however, the problem of spatial autocorrelation may not be a serious one essentially because here the sample points, being cities and residential and commercial centres, are mostly not contiguous and are scattered over the country concerned. Hence, any change in the atmospheric air quality recorded for one city/centre is unlikely to affect significantly that of another in the sample.

As a part of the preliminary data analysis, we have examined the summary statistics relating to the pollution data (i.e., mean, variance and correlation coefficient with *PCGDP*). These are reported in Table 4.2. It should be noted that the average *SPM* level for residential areas is higher than that for commercial areas, but the mean *PCGDP* level is higher for commercial areas than that for residential areas. This is possibly because of the fact that the residential areas in the present data set are mostly located in the less developed and developing countries. The correlation coefficient between *SPM* and real *PCGDP* may be seen to be negative and large separately for each data set and also for the combined data sets. The smallest absolute value of this correlation is 0.79. It may be noted that this contradicts the *EKC* hypothesis. However, such contradictory empirical results have been obtained earlier also. For example, Grossman and Krueger (1993, 1995) and Torras and Boyce (1998) reported results not supporting the *EKC* hypothesis for ambient *SPM* and heavier particles, respectively[39]. This is confirmed if we look at the scatter diagrams, all of which show the same decreasing pattern (Figure 4.1). A possible explanation of this may be that the present data set contains observations relating to mostly developed countries (which may have crossed the so-called turning point of the *EKC*). Table 4.3, which gives the distribution of countries by selected level of *PCGDP* (assumed to correspond to the possible turning point of the *EKC*), may corroborate this. Thus, e.g., if the level of *PCGDP* corresponding to the turning point of the *EKC* for *SPM* is taken to be $8,000 (the result of Selden and Song (1994), then 20 out of the 34 sample observations would belong to the declining portion of an inverted U-shaped *EKC* for *SPM*.

Table 4.3. Distribution of countries by the level of *PCGDP* corresponding to turning point of *EKC*

Pollutants	Group	0 - $ 3000	$3000- $6000	$6000-$8000	$8000 &more
SPM	c	6	7	1	20
	r	8	2	0	4
	All	14	9	1	24
SO_2	c	8	14	7	20
	r	5	10	2	3
	All	13	24	9	23

Table 4.4. Groupwise results of OLS regression for SPM data

Group	Estimated Coefficients of Explanatory Variable (income)			R^2 (d.f.)	$\sqrt{\sum e^2/n}$	$\sum \lvert e \rvert /n$	Turning points
	Intercept	x	x^2				
c_1	419.43*** (14.15)	-0.073*** (-6.54)	.36E-5*** (4.08)	0.93 (11)	32.30	25.37	10033
c_2	437.00*** (10.4)	-0.067*** (-4.81)	.29E-5** (2.91)	0.89 (10)	41.31	27.70	11538
c_3	466.76*** (7.38)	-0.065** (-3.75)	.25E-5** (2.79)	0.91 (4)	44.13	26.70	13238
c	418.68*** (20.6)	-0.060*** (-11.57)	.24E-5*** (7.84)	0.89 (31)	38.17	29.78	12618
r_1	382.06*** (6.72)	-0.090* (-2.48)	.5E-5 (1.9)	0.89 (3)	55.50	28.42	8374
r_2	375.59** (3.75)	-0.067 (-1.22)	.33E-5 (0.84)	0.76 (3)	87.16	44.92	10225
r	371.37*** (8.13)	-0.070** (-2.74)	.37E-5* (1.9)	0.80 (11)	61.93	39.60	9529
All	382.07*** (19.37)	-0.055*** (-9.11)	.22E-5*** (5.75)	0.81 (45)	52.89	37.92	12500

Note: Figures in parentheses are the t-ratios. Pollution is measured in mg/m^3. Income is measured in terms of 1985 US dollars. One, two and three asterisks indicate that a coefficient estimate is significantly different from zero at 10%, 5% and 1% level, respectively.

39 Grossman and Krueger (1995), however, did not find a minimum point of the estimated curve for heavy particles.

Tables 4.4 – 4.6 present our regression results for *SPM*. The scatter diagrams in Figure 4.1A suggest that the shape of the underlying relationship between *PCGDP* and *SPM* is U-shaped. The ordinary least squares (OLS) estimates of the corresponding (equation 4.1) quadratic relationship between *PCGDP* and *SPM* for different periods and areas are reported in Table 4.4. All these results show a negative estimated value of β_1 and a positive estimated value of β_2, both being statistically significant[40]. Thus, for *SPM* our results suggest a U-shaped relationship between *SPM* and *PCGDP*, which implies that beyond a certain level of *PCGDP* (around \$12,500), a further rise of *PCGDP* can be achieved at the cost of environmental degradation[41]. Clearly, this result contradicts the usual *EKC* hypothesis, but supports some earlier findings. For example, Kaufmann et al., (1998) found a U-shaped relationship between income and atmospheric concentration of SO_2 with a turning point around the *PCGDP* level of \$12,000. Sengupta (1997) also noted that beyond the per capita income \$15,300 the environmental base (particularly CO_2 emission) *relinks with* economic growth. Finally, Shafik (1994) obtained upward rising curves by fitting cubic relationships.

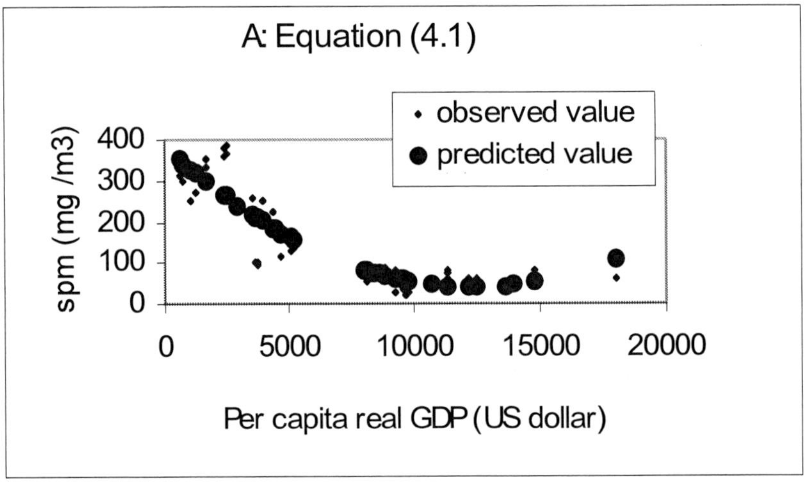

Figure 4.1. Continued on next page.

---

[40] β_1 and β_2 are the coefficients of equation (4.1).

[41] An alternative measurement also reveals the same result. Instead of PCGDP, we took Gross City Product Per Capita (GCPPC) from World Resources 1998 – 99 (World Resources Institute et al., (1998)). Using GCPPC and *SPM* (mg/m^3) for the year 1993, we found the same result, viz., U-shaped relationship between SPM and GCPPC. This later data set covered 22 cities across the world. The estimated relationship is:

SPM = 215.4 – 0.01906(GCPPC) +0.416E-6 (GCPCDP)^2.

(7.5) (-2.755) (2.001)

and the coefficients of GCPPC and square of GCPPC are significant at 5% and 10% level, respectively. In case of SO_2, after removing an outlier, we obtained a negatively sloped linear relationship.

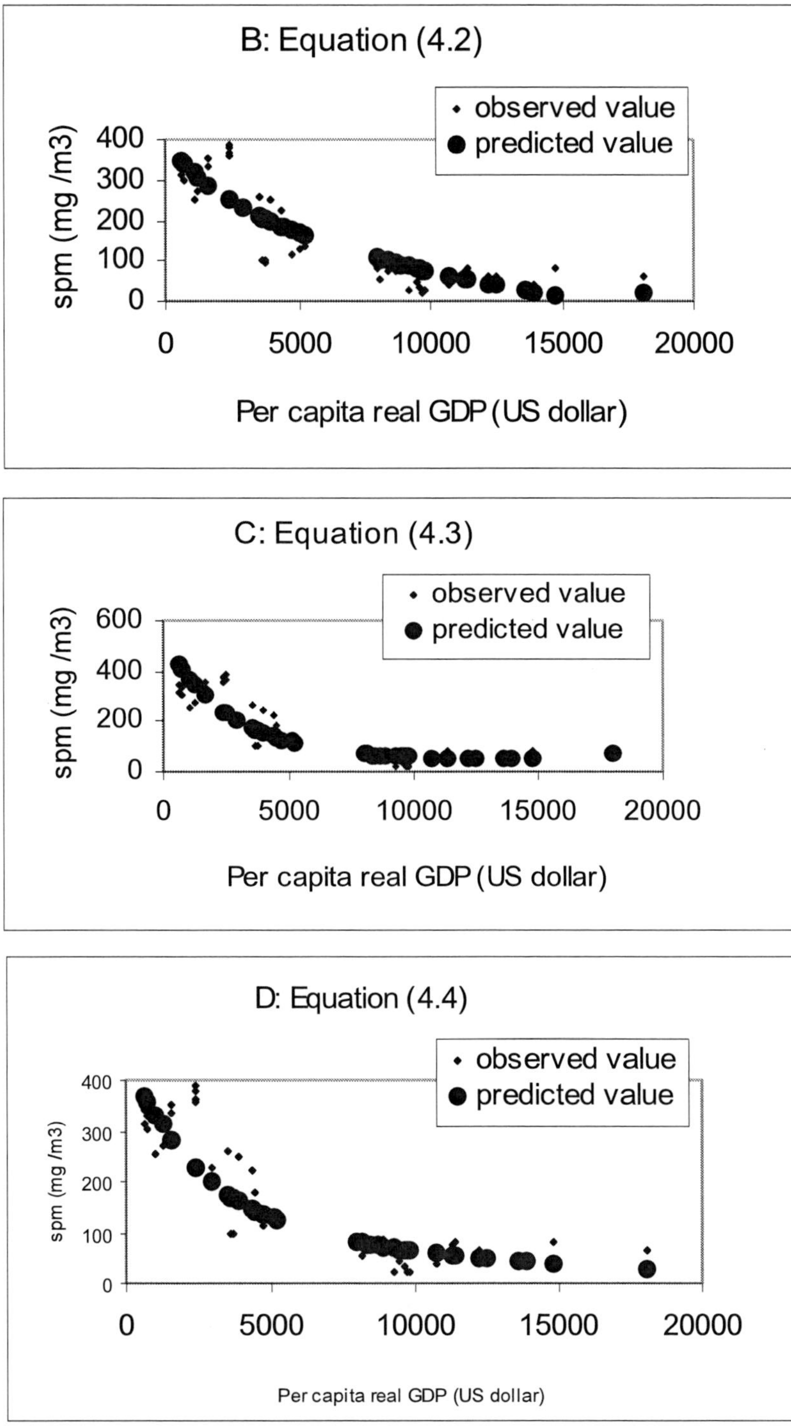

Figure 4.1. Relationship between PCGDP and SPM.

Our OLS diagnostic test identified one outlier in the present data set. To take care of this, we re-estimated all the regression specifications using the Least Absolute Error (LAE) method[42]. The estimated LAE and OLS results are presented in Table 4.5. As is to be expected, the estimated LAE results are very similar to the corresponding OLS results. Perhaps the most interesting findings for *SPM* again are the U-shaped relationship with rather high *PCGDP* values corresponding to the turning point (*vide* last columns of Tables 4.4 and 4.5). This is in contrast to the results of Selden and Song (1994) and Grossman and Krueger (1995), who observed turning point for *SPM* around *PCGDP* levels of $8,000 and $5,000, respectively. To be precise, our turning point estimates for *SPM* vary between $ 9,500 and $ 14,000.

Table 4.5. OLS & LAE regression results of SPM on *PCGDP* for different forms separately for commercial and residential areas

Class	OLS LAE	Estimated Coefficients of Explanatory Variable					R^2 (d.f.)	Turning points
		Intercept	x	x^2	P_1	P_2		
c	OLS	418.68***	-0.06***	0.24E-5***			0.89 (31)	12618
		(20.62)	(-11.57)	(7.84)				
	LAE	373.17	-0.053	0.21E-5			0.87	12604
	OLS	430.42***	-.058***	0.22E-5***	-27.24	-12.22	0.90 (29)	13122
		(18.66)	(-10.17)	(6.31)	(-1.31)	(-0.59)		
	LAE	405.59	-0.05	0.17E-5	-34.07	-25.44	0.89	14504
r	OLS	371.37***	-0.07**	0.37E-5*			0.80 (11)	9529
		(8.13)	(-2.74)	(1.94)				
	OLS	374.43***	-0.07**	0.37E-5	-6.94		0.80 (10)	9559
		(7.32)	(-2.56)	(1.8)	(-0.18)			
	OLS	395.02***	-0.079**	0.42E-5	-31.63	0.005	0.81 (9)	9524
		(6.06)	(-2.44)	(1.8)	(-0.53)	(0.55)		
All	OLS	382.07***	-.055***	0.22E-5***			0.81 (45)	12500
		(19.37)	(-9.11)	(5.75)				
	LAE	368.75	-0.052	0.2E-5			0.81	12667
	OLS	387.77***	-.054***	0.21E-5***	-16.67		0.82 (44)	12717
		(18.98)	(-8.88)	(5.47)	(-1.05)			
	OLS	395.77***	-.053***	0.2E-5***	-27.13	-14.25	0.82 (43)	12998
		(16.28)	(-8.26)	(4.9)	(-1.17)	(0.62)		

Note: Figures in parentheses are the t- ratios. Pollution is measured in mg/m^3. Income is measured in terms of 1985 US dollars. One, two and three asterisks indicate that a coefficient estimate is significantly different from zero at 10%, 5% and 1% level, respectively.

Table 4.5 presents the estimated OLS and LAE results for commercial, residential areas separately and also for the combined data for the two types of areas. So far as these estimates are concerned, it should be noted that the OLS and the corresponding LAE estimates are broadly similar, both in terms of goodness of fit and magnitude of the estimated parameters (however, unique LAE estimate could not be obtained in some cases). A closer look at Table 4.5 may suggest the following results. First, the values of R^2 and the *PCGDP* corresponding to the turning point for residential areas are smaller than those estimated for commercial areas. Next, while the estimated coefficients of *PCGDP* (i.e., β_1) are negative and those of

[42] In Econometric theory, LAE estimates are regarded as robust estimates in non-normal data situations containing some extreme observed values. See Judge et al., (1985) Chapter 20 for the LAE estimation method and the properties of the LAE estimator.

square of *PCGDP* (i.e., β_2) are positive in all the cases, the estimated β_2 coefficients for residential areas are not highly significant. Thus, statistically speaking, the U-shape of the Pollution-PCGDP relationship is weaker for the data relating to the residential areas, but is rather strong for the data relating to the commercial areas. The estimated values of *PCGDP* corresponding to the turning point are estimated to be around $ 9,500 and $ 12,500 for residential and commercial areas, respectively[43]. Interestingly, the high-income countries observed to lie beyond the turning point in the present exercise included the USA, Canada, Japan, Finland and Germany. One might seek an explanation of difference in the results for the two types of areas in terms of how the relative density of population in these two types of areas changes with economic growth.

In the next part of the exercise an attempt is made to have a causal explanation of the observed U-shaped / inverse Pollution-PCGDP relationships. *A priori*, one should expect the pollution level in an economy to depend not only on the level of *PCGDP*, but also on the sectoral composition of GDP, how the *PCGDP* level is being achieved, and the time rate of growth of *PCGDP*. The sectoral composition is important, because *ceteris paribus* an economy with a larger industrial production is likely to have more pollution. The nature of the production technique used may be relevant, because often a more capital-intensive production technique is likely to be more non-human energy-intensive and hence more polluting. Finally, the rate of growth of *PCGDP* may be a determining factor since *ceteris paribus* a faster growth may commonly be achieved by exercising the softer option of using more polluting production practices. In other words, a strong urge to grow faster, given the level of *PCGDP*, may induce a less developed economy to adopt a less clean production technique. Coming to the possible partial effect of production technique (as represented by the capital-labour ratio) of an economy, say, it may be argued that between two countries with the same level of *PCGDP*, one having a greater concern for pollution would have a higher capital-labour ratio, if a cleaner technology is more capital intensive[44]. Thus, we tried to examine the validity of the following hypotheses - (i) the marginal change in pollution level with respect to *PCGDP* is increasing in the rate of growth of *PCGDP* and decreasing in time; and (ii) the marginal change in pollution level with respect to *PCGDP* is decreasing in both the capital- labour ratio and the sectoral composition of GDP. To examine the possible partial effects of

---

[43] These figures are higher than those found in the studies of Selden and Song (1994), Grossman and Krueger (1995), Shafik (1994), World Bank (1992). Kaufmann et al., (1998), on the other hand, found a U-turn for the atmospheric concentration of SO_2 at *PCGDP* level around $ 12,000. Grossman and Krueger (1995) also observed an upswing of the pollution level at about a *PCGDP* level of $ 16,000. However, since there were only two observations beyond these levels, existence of such a reverse upswing at high level of *PCGDP* was not claimed.

[44] A cleaner industrial technology would frequently be more expensive and hence more capital intensive because of the technical sophistications involved – take, e.g., the catalytic converters used to reduce lead emission from automobiles. There may however be innovation leading to less polluting and at the same time less expensive production techniques, but such innovation is infrequent. A country having catalytic converters is more capital intensive with lower emission than another not having these, *ceteris paribus*. It should be noted that capital abundant countries have an option to use a part of capital for abatement activity and a country investing in such abatement activity will have less pollution. From this point of view, a larger capital-labour ratio need not necessarily mean greater pollution. Note that we have not considered here the issues of trade. However, if the two countries are open to trade, the Heckscher-Ohlin theorem predicts that capital-intensive industries locate in capital-abundant countries. Thus, when there are no differences in per capita income that could lead to differences in the technique effect, the country with the lower capital abundance will have lower emissions.

production technique[45] and sectoral composition of GDP on pollution, the following regression set up is used:

$$y_{it} = \beta_0 + \beta_1 x_{it} + \gamma_1 p_1 + \gamma_2 p_2 + \delta_1 z_1 + \delta_2 z_2 + \eta_1 d_1 + \eta_2 d_2 + \theta_1 w_1 + \theta_2 w_2 + e_{it} \quad (4.5)$$

where y_{it} and x_{it} are as already defined, p_j: dummy variable representing time period (viz., $p_1 = 1$ for the period 1979-1982 and zero otherwise, $p_2 = 1$ for the period 1983-1986 and zero otherwise); d_1: dummy variable for capital intensity (viz., $d_1 = 1$ for a country having capital-labour ratio greater than or equal to 1 and zero otherwise); d_2 : dummy variable for share of non-agricultural sector in GDP (viz., $d_2 = 1$ for a country for which the non-agricultural sector accounts for 90 per cent or more of GDP and zero otherwise); $z_j = x * p_j$, j=1,2 are the income-time period interaction terms; $w_1 = x * d_1$ is the income-capital intensity interaction term; $w_2 = x * d_2$ is the income-share of non-agricultural sector interaction term; and e_{it} is the equation disturbance term.

Let us first discuss the results relating to the effects of capital intensity and sectoral composition of GDP on the pollution level. Table 4.6 presents these results for *SPM.* So far as the level of *SPM* (i.e., the intercept term of the regression of *SPM* level on *PCGDP*) is concerned, in none of the equations the coefficients of the time dummy variables were statistically significant implying thereby that the level of *SPM* did not shift perceptibly over time. As regards the effect of capital intensity on the *SPM* level (i.e., the intercept dummies for this variable), this was observed to be negative and highly significant for the data relating to the residential areas and the combined data, but non-significant for the data relating to the commercial areas. A similar significant negative level effect of the sectoral composition variable was also observed for all the three data sets.

Let us next describe the results showing how the marginal change in pollution in response to a change in the level of *PCGDP* (i.e., the slope term of the regression of *SPM* level on *PCGDP*) are affected by the time dummy, capital intensity and the sectoral composition variables. These are given by the estimated values of the parameters associated with the interaction terms of *PCGDP* and these variables (viz., the values of the parameters δ_1 and δ_2 measuring the effect of interaction between time and *PCGDP*, θ_1 measuring the effect of interaction between capital intensity and *PCGDP*, and θ_2 measuring the effect of interaction between sectoral composition of GDP and *PCGDP*, respectively in equation (4.5)). The interaction effect between time and *PCGDP* is negative and significant only for the data relating to the commercial areas. This implies that compared to 1979-82 in latter periods the

[45] In our empirical analysis reported in this chapter we have tried to explain the level of pollution in terms of production technique (as reflected by the capital-labour ratio for the economy as a whole) and sectoral composition of GDP of individual countries, in addition to *PCGDP*. Country-wise capital-labour ratios have been calculated on the basis of country-wise data on gross capital and employed labour force available in UN's

decrease in pollution in response to a marginal increase in *PCGDP* was greater. Next, the interaction effect between capital intensity and *PCGDP* is positive and highly significant for the data relating to the commercial areas and also for the combined data. For the data relating to the residential areas this effect is however negative and significant at 10 per cent level. The positive interaction effect suggests that, *ceteris paribus*, a country with a higher capital intensity would have a lower, but flatter, Pollution- *PCGDP* curve compared to one with a lower capital intensity[46].

Table 4.6. OLS regression results of SPM on GDP and different dummy variables for different groups and their combination

Group	Estimated Coefficient of Explanatory Variable										$\bar{R}^2$
	Intercept	x	P_1	P_2	Z_1	Z_2	d_1	d_2	W_1	W_2	
c	318.4***	-0.018***	23.65	23.64	-0.012**	-0.01*					0.78
	(7.54)	(-4.68)	(0.45)	(o.43)	(-2.24)	(-1.77)					
	311.5***	-0.02***					-32.36				0.71
	(14.06)	(-5.82)					(-1.08)				
	356.2***	-0.03***					-209.4***		0.02***		0.80
	(16.5)	(-7.87)					(-4.07)		(3.92)		
	418.7***	-0.051***						-320***		0.047***	0.90
	(18.35)	(-7.38)						(-8.19)		(6.29)	
r	474.3**	-0.07*	-262.67	-227.5	0.054	0.053					0.21
	(2.7)	(-1.92)	(-1.3)	(-1.12)	(1.4)	(1.39)					
	340.2***	-0.0004					-282.7***				0.92
	(16.67)	(-0.14)					(-9.37)				
	281.5***	0.037*					-218.08***		-0.04*		0.93
	(7.77)	(1.84)					(-4.97)		(1.88)		
	377.9***	-0.057***						-296**		0.05**	0.59
	(6.4)	(-3.437)						(-2.96)		(2.92)	
All	288.1***	-0.017***	12.6	7.28	-0.008	-0.006					0.56
	(6.16)	(-3.47)	(0.21)	(0.12)	(-1.21)	(-0.83)					
	297.2***	-0.014***					-107.42***				0.67
	(15.3)	(-4.57)					(-4.12)				
	363.8***	-0.031***					-273.26***		0.03***		0.82
	(20.15)	(-8.62)					(-8.2)		(6.13)		
	406.8***	-0.054***						-317***		0.05***	0.80
	(16.23)	(-7.3)						(-7.39)		(6.28)	

Note: Pollution is measured in mg/m^3. Income is measured in terms of 1985 US dollars. Figures in parentheses are the t- ratios. One, two and three asterisks indicate that a coefficient estimate is significantly different from zero at 10%, 5% and 1% level, respectively.

National Accounts Statistics and ILO's Yearbook of Labour Statistics, respectively. Finally, country-wise data on sectoral composition of GDP have been obtained from the World Bank reports.

[46] These results are not very strong as we have used a dummy variable to represent the capital abundance. Use of continuous cross section data on capital abundance for different countries in lieu of the dummy variable may give different results.

Table 4.7. OLS & LAE regression results of SO_2 on GDP

Group	OLS LAE	Estimated Coefficient of Explanatory Variable						R^2 (d.f.)	$\sqrt{\sum e^2/n}$	$\sum\lvert e\rvert/n$
		Intercept	x	x^2	P_1	P_2	c			
c	OLS	78.53*** (9.4)	-0.0039*** (-3.65)					0.220 (47)	27.96	22.86
	OLS	71.6*** (5.49)	-0.0038*** (-3.43)		7.06 (0.61)	7.75 (0.67)		0.230 (45)	28.42	23.06
	LAE	62.43	-0.0028					0.130	30.39	22.45
	OLS	80.64*** (6.3)	-0.0047 (-1.35)	0.5E-7 (0.22)				0.220 (46)	28.25	22.72
	LAE	81.9	-0.0063	0.14E-7				0.200	35.41	32.95
r	OLS	48.93*** (3.76)	0.0005 (0.245)					0.003 (18)	31.98	24.8
	OLS	50.72 (2.17)	-0.00025 (-0.03)	0.6E-7 (0.09)				0.004 (17)	32.89	24.65
All	OLS	68.7*** (9.75)	-0.0027*** (-2.79)					0.100 (67)	29.63	24.65
	OLS	59.15*** (5.14)	-0.0028*** (-2.735)		5.9 (0.58)	9.86 (0.98)	4.95 (0.6)	0.122 (64)	29.99	24.65
	LAE	56.09	-0.0023						32.88	23.25
	OLS	66.99*** (6.15)	-0.002 (-0.65)	-0.4E-7 (-0.207)	6.07 (0.57)	10.02 (0.95)		0.100 (64)	29.84	24.65
	OLS	59.5*** (4.54)	-0.003 (-0.88)	0.13E-7 (0.06)			4.98 (0.6)	0.123 (65)	30.23	24.37
	LAE	54.05	-0.002	-0.2E-7					28.31	23.24

Note: Pollution (SO_2) is measured in mg / m^3. Income is measured in terms of 1985 us dollars. Fssigures in parentheses are the t- ratios. One, two and three asterisks indicate that a coefficient estimate is significantly different from zero at 10%, 5% and 1% level, respectively.

Finally, the interaction effect between sectoral composition of GDP and *PCGDP* was estimated to be positive and significant for all the three data sets. This, together with the fact that the coefficient of the corresponding intercept dummy is negative and significant, suggests that, *ceteris paribus*, more industrialised countries have a lower, but flatter, Pollution-PCGDP curve.

In this context it may be mentioned that for all the data sets the goodness of fit of the quadratic Pollution-PCGDP equation is more or less similar to those of the corresponding regression equation in which *PCGDP*, capital intensity (sectoral composition of GDP) and interaction between *PCGDP* and capital intensity (sectoral composition of GDP) are used as separate regressors. This possibly means that in association with *PCGDP* structural factors like production technique and sectoral composition may help explain observed changes in pollution level over time or across region. In other words, the quadratic term on the r.h.s. of equation (4.1) is in fact replaced by $(\gamma_1 p_1 + \gamma_2 p_2 + \delta_1 z_1 + \delta_2 z_2 + \eta_1 d_1 + \eta_2 d_2 + \theta_1 w_1 + \theta_2 w_2)$ to yield equation (4.5), because *a priori* rate of change of marginal pollution due to *PCGDP* level may be due to total effects of technology, sectoral composition of GDP, time and their interaction with income level.

The results of the analysis of our SO_2 pollution data are summarised in Table 4.7. Compared to the analysis of the *SPM* data, fewer interesting findings are obtained in this case. To be precise, while the correlation coefficients between SO_2 and *PCGDP* were observed to be negative for data sets relating to commercial areas, the corresponding correlation coefficients for residential areas were observed to be positive. This may be due to the fact that the data set for residential areas included data for only three developed countries, viz., USA, Canada and the New Zealand, whereas the data set for commercial areas covered, in addition to these three countries, a number of other developed countries.

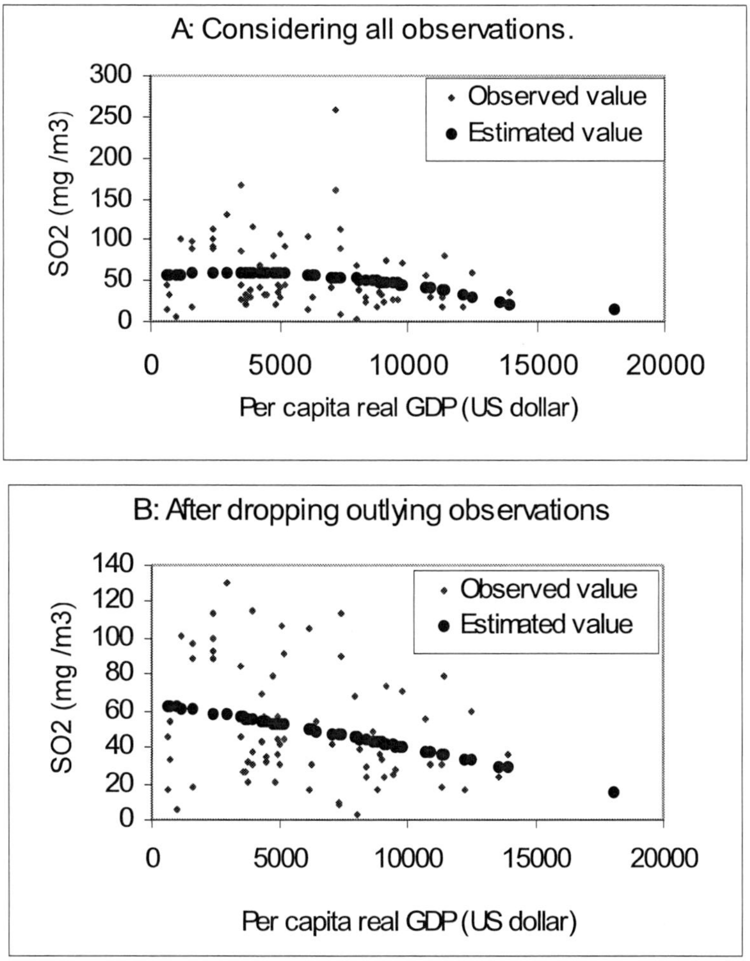

Figure 4.2. Relationship between PCGDP and SO_2.

Examination of the scattered diagrams suggested wide variation of the SO_2 level at low level of *PCGDP* that gradually narrowed down as the *PCGDP* level increased. Further probe suggested presence of some outliers (viz., data relating to Iran and Italy) in the data set, which were dropped in subsequent analyses[47]. Removal of these out-liers resulted in a linear

[47] The data for Iran was unusual possibly because of the Iraq-Iran war during 1977-88, whereas Italy experienced a series of volcanic eruptions during the early 1980s.

relationship with a negative slope (not an inverted U-shaped relationship) between SO_2 level and *PCGDP* (Figure 4.2b). These results thus suggest absence of any clear relationship between the level of SO_2 and *PCGDP* for data relating to the residential areas. A possible explanation of the observed relationship for commercial areas could be that the extent and the quality of automobile emission[48] improved considerably with rise in *PCGDP*. In addition, the type of fuel used for domestic and commercial purposes in low income developing countries might contribute to the relatively high level of atmospheric SO_2 in them. With economic growth a transitional forces strengthen the market mechanism and as a result the economy gradually shifts from non-commercial to commercial energy resources. There may also be other reason – viz., high-income countries tend to spend more on defensive expenditure, enforce a stricter environmental regulation and use cleaner technology which others can not afford.

4.4. Conclusion

The basic objective of the present study was to re-examine the hypothesis of EKC using cross-country time series data on two air pollutants, viz., SPM and SO_2. Our results do not support the EKC hypothesis. In contrast, for SO_2 we obtained an inverse relationship with PCGDP, while for SPM a U-shaped, rather than an inverted U-shaped, relationship with PCGDP is observed with an upward turn of the curve around a PCGDP level of $ 12,500 which represents a rather high level of material consumption. To the extent the level of currently available technology is unable to ensure sustainability of such a high consumption level, a further rise of PCGDP beyond the threshold level can support consumption only at the cost of a slow but steady deterioration of the environmental quality.

To explain the observed Pollution-PCGDP relationship, three economic variables other than PCGDP were brought into the analysis - viz., the economy-level capital intensity, the sectoral composition of GDP, and the rate of growth of GDP. It was thought that given the PCGDP level of an economy, these three aspects would determine the exact nature of relationship that might exist between pollution and income level. In other words, it is not only the level of income but also the characteristics of an economy which together determine the rate of environmental degradation that an economy will experience as it moves along the trajectory of development. Although the way these variables have been used in the present study leaves scope for improvement, their inclusion does give meaningful and statistically significant results so far as the explanation of the phenomenon of pollution is concerned. Briefly, our results suggest that the partial effect of capital intensity on pollution is generally negative (which may not be unreasonable, if the trend of technological progress is such that more capital-intensive techniques are more environment-friendly and vice versa). The observed negative partial effect of the sectoral composition variable on pollution perhaps suggests that, given the PCGDP level, the more industrialised an economy is the lower and flatter would be its Pollution-PCGDP curve. Finally, PCGDP and the rate of growth variable seem to be jointly important in explaining observed pollution level of an economy.

[48] See, Kahn (1998).

This chapter examined the presence or otherwise of significant statistical association between the level of economic activity and environmental degradation without explicitly discussing the nature of causation between them. In chapter 5 and 6 of this book, we shall concentrate on the causality aspect.

Chapter 5

CAUSALITY BETWEEN INCOME AND EMISSION: COUNTRY-GROUP SPECIFIC ECONOMETRIC ANALYSIS*

5.1. INTRODUCTION

Concerned with the problem of increasing environmental degradation, researchers have been examining the nature of relationship between the level of economic activity or income and environmental quality indicators. In recent period there have been extensive empirical studies to testify the Environmental Kuznets Curve (EKC) hypothesis which postulates *an inverted U-shaped relationship between level of economic activity and environmental pressure* (defined as *the level of concentration of pollution or flow of emissions, depletion of resources etc.*)[49]. The literature has mostly considered EKC as an empirical phenomenon and examined the presence or otherwise of significant statistical association between the level of economic activity and environmental degradation without explicitly discussing the nature of causation between these variables. Using cross-country cross-sectional data, these empirical studies estimated a regression equation of some measure of environmental degradation on some measure of level of economic activity (like per capita GDP or income) and examined whether or not the underlying true regression relationship might be of the inverted u-shape. These exercises thus presume a unidirectional causal relationship – viz., a change in the level of economic activity/per capita income causes a consequent change in the environmental quality. While this presumption appears reasonable, it may not hold in all circumstances. In fact, as we shall explain later, whether a change in the level of economic activity would cause or would be caused by a change in the environmental quality should depend upon various characteristics of the economy under consideration.

Given a time series data set on measures of level of economic activity and corresponding environmental change, one may use time series econometric techniques like Granger

---

* This chapter is based on Coondoo and Dinda (2002).

[49] See, e.g., Grossman and Krueger (1995), Suri and Chapman (1998), Selden and Song (1994), Panayatou (1997), Carson et al., (1997), Kahn (1998), McConnell (1997), Torras and Boyce (1998), List and Gallet (1999), Koop and Tole (1999), for EKC studies using pollution variables other than CO_2 emission and Holtz-Eakin and

Causality Test (GCT) to examine whether a statistically significant causality exists between the two variables and if so, what is the direction of causality. As we shall explain later, causality here is understood in a very specific and well-defined sense (Hamilton, 1994). A number of empirical studies examining such causality have already been done[50].

The studies referred to above provide ample evidences of the fact that the direction of causality between the level of income and environmental quality need not always be unidirectional from income to environmental quality, as usually thought. In fact, the presumption of a direction of causality may hamper a fuller understanding of the true nature of the environment-income relationship[51]. Further, as the direction of causation between environmental quality and economic growth has significant policy implications, any such presumption may lead to erroneous policy conclusions[52]. The empirical results obtained so far seem to reinforce the need to make deeper probe into the causality aspect of the income-environment relationship. The present study makes such an attempt. Using a panel data set on per capita income and per capita CO_2 emission for a large number of countries spread all over the Globe, we have tried to do a careful statistical analysis primarily using the GCT. To make the investigation comprehensive and exhaustive, we have examined presence or otherwise of a causal relation between income and CO_2 emissions for groups of similar countries in different continents, using the cross-country panel data set and also the corresponding aggregate time series data for the country groups. As to be expected, a variety of results obtained showing different directions of causality for different groups of countries/continents at different level of aggregation.

This chapter is organized as follows. Section 5.2 discusses the issue of causality in the context of EKC from an economic theoretic standpoint and explains how this links up with the concept of causality underlying the GCT. Section 5.3 explains the sources and the nature of the data used. Section 5.4 presents the main empirical results. Finally, Section 5.5 draws some concluding observations. The econometric methodology of GCT followed in this study is explained in the Appendix B and a supplementary information is presented in Table A.1 in Appendix A.

5.2. Causality in the Context of *EKC* Hypothesis and *GCT*

As mentioned earlier, the *EKC* hypothesis postulates an inverted u-shaped relationship between environmental degradation/pollution and the level of economic activity/per capita income of an economy, with the former taken as the dependent variable of the postulated relationship. For convenience of exposition, in what follows we shall call levels of economic

Selden (1995), Shafik (1994), Sengupta (1997), Cole et al., (1997), Moomaw and Unruh (1997), de Bruyn et al., (1998), Unruh and Moomaw (1998) for EKC studies using CO_2 emission as the pollution variable.

50 See, e.g., Yang (2000), Glasure and Lee (1997), Cheng (1996,1999).

51 In some studies of EKC whether income level is at all an important determinant of environmental quality has been examined. See, e.g., Agras and Chapman (1999), who observe that when energy price and trade related variables are used as explanatory variables along with income, it is the energy price and not the income which becomes the significant determinant of environmental quality (i.e., CO_2 emission) or energy demand. In a way thus the Agras and Chapman study raises a question about the desirability of presuming any causal relationship between environmental quality and income level.

52 See, Goulder and Schneider (1999).

activity and environmental degradation as income and emission, respectively[53]. The *EKC* hypothesis actually summarizes *an essentially dynamic process of change* – viz., as income of an economy grows over time, emission level grows first, reaches a peak and then starts declining after a threshold level of income has been crossed. However, the statement of the hypothesis makes no explicit reference to time. Evidently thus, under the null hypothesis of *EKC* and under the assumption of *invariance of the emission-income relationship*[54], for a given set of cross-country cross sectional data on income and emission, the *emission on income* regression line should be an inverted U-shaped empirical *EKC*.

Insofar as the cause-effect relationship between emission and income is concerned, two obvious alternative representations are *emission = f (income)* and *income = g (emission)*, *f (.)* and *g (.)* denoting the functional form of the relationships. One may interpret the first one as the *engel curve* for emission (which is typically regarded as a *bad* item from the point of view of *consumer preferences*). Under this interpretation, the *EKC* hypothesis would mean that the income elasticity of emission declined to zero with rise in income and became negative beyond the threshold income level (i.e., the income level at which income elasticity had became zero). In other words, under the *EKC* hypothesis, with growth of income the status of emission *as an item of consumption* gradually changes from a *necessary* to an *inferior good* (thus reflecting a clear preference for a cleaner environment at higher levels of living).

The second relationship regards emission as the *cause* and income as the *effect* variable. This may be given a production relation interpretation - viz., emission is an *essential* input for income generation and without emission income generation is impossible[55]. Juxtaposition of this interpretation with the *EKC* hypothesis suggests two distinctly different production regimes. The first one corresponds to income levels lower than the threshold income. In this regime income growth requires rising emission. The second one corresponds to income levels above the threshold income. This is characterized by declining emission with income growth. It may be noted that a positive association between income and emission (that characterizes the first regime) was the common experience of all the developed nations of today's world during the stage of their industrialization. As regards the second regime, this may materialize under alternative conditions. For example, a country may experience income growth with a corresponding decline in emission, if composition of its GDP gradually changes away from emission-intensive manufacturing to less emission-intensive services (as it is believed to have been happening in the developed economies like that of the U.S.A., say)[56]. Alternatively, if in course of development a country is able to substitute conventional fossil fuel by alternative energy resources having less emission (i.e., makes a conscious effort to improve environment), then also a negative association between emission and income may emerge, even if the country's income continues to be manufacturing-intensive[57].

---

[53] To ensure comparability across countries with different population sizes, these variables should be normalized by the population and one should use per capita income and per capita emission as the basic variables defining the *EKC*.

[54] That is, *ceteris paribus*, in their process of development individual countries experience income and emission situations lying on *one and the same EKC*.

[55] This kind of causal direction is referred to in the literature as *reverse causality*.

[56] International trade may facilitate such a compositional change of GDP of the developed countries as these countries may shift the production of emission-intensive manufactured items shifted to the developing world and get these imported.

[57] Obviously, thus, the reverse causality would be a serious issue and matter of concern for countries using conventional fuel for which the share of manufacturing in the GNP continued to be high as income increased over time.

To interpret the *EKC* hypothesis, one may combine the above two types of causation as well. Thus, it may be said that the *emission to income* causation works in the rising portion of the *EKC* corresponding to lower levels of income. Once the threshold income level has been achieved, a strong societal demand for a cleaner environment forces a gradual shift to less emitting production technology. Thus, the *income to emission* causation may be said to work in the falling portion of *EKC* when the hypothesis holds.

The above discussion, however, is inadequate on at least two counts. First, it takes a partial view of the effect of emission/environment either from the point of view of consumption or from the point of view of production/income generation. Such a partial view is over-simplistic, because, as is well recognized, *a priori* emission or environment may affect both consumers' welfare (as a non-excludable public good) and income generation (by virtue of being a virtual input to the income generation process[58]). More importantly, it presumes an immediacy in the causal relationship (i.e., as if a change in one of the variables would instantaneously cause the other to change) and hence does not clearly bring out the dynamic process of changes that is so crucial in the *EKC* relationship. To explain this dynamic process, let us consider the following simple inter-temporal choice model.

Consider a one-good economy for which environment E, understood as a stock variable, affects both utility and production level. Let $C(t)$, $E(t)$ and $K(t)$ denote consumption, environment and capital stock at time t. Let us assume that $\theta(t)$ $(0<\theta(t)<1)$ portion of capital stock is used for production of the good and the remaining $(1-\theta(t))$ portion is used for upgrading the environment. Finally, let $\gamma\,(>0)$ be the rate of pollution

(i.e., emission or degradation of environment per unit of output produced (for detail see Dinda (2005)). The infinite time horizon inter-temporal choice problem may be specified as

$$\text{Maximize } W = \int_0^\infty e^{-\rho t} U(C(t), E(t))dt \tag{5.1}$$

Subject to the accumulation constraints

$$\dot{K}(t) = f(\theta(t)K(t), E(t)) - C(t) \tag{5.2}$$

and

$$\dot{E}(t) = g((1-\theta(t))K(t)) - \gamma f(\theta(t)K(t), E(t)) \tag{5.3}$$

Where ρ is the discount rate and $f(.)$ and $g(.)$ are the production function and the *environment upgrading* function of the economy[59]. Clearly, the first constraint relates to physical capital formation while the second constraint relates to the net environmental change

---

[58] What is meant here is essentially the sustainability argument – i.e., emission/environmental degradation may have strong negative effect on output/income generation such that in the long run the income generation process may collapse.

[59] This model is based on Dinda (2005) and all first order conditions are available in its appendix.

due to production and environmental upgrading[60]. Treating *C(t)* and $\theta(t)$ as control variables and *K(t)* and *E(t)* as state variables and assuming usual regularity conditions for the production and the utility function, the optimality condition for the above problem turns out to be

$$\alpha(t)\frac{\dot{C}(t)}{C(t)}+\beta(t)\frac{\dot{E}(t)}{E(t)}=\phi(t) \tag{4.4}$$

Where $\alpha(t)=\dfrac{C(t)U_{CC}}{U_C}$, $\beta(t)=\dfrac{E(t)U_{CE}}{U_C}$ and $\phi(t)=(-\dfrac{f_K g_K}{g_K+\gamma f_K}+\rho)$,

U_C, U_{CC}, U_{CE} being the relevant first and second order partial derivatives of $U(.)$ and f_K and g_K being the first order partial derivative of the functions $f(.)$ and $g(.)$ with respect to *K*. Note that the equation (5.4) suggests that time paths of income (i.e., *C* here) and environment (i.e., *E* here) should generally be interdependent. This, thus, means a two-way causal relationship between income and environment, in general. Next, consider the special case where $U_{CC}=0$, but $U_{CE}\neq 0$. In this case the rate of change of environmental stock will depend on the capital accumulation or consumption/income. This suggests that given an autonomously chosen time path of income, the corresponding time path of environment will be determined by the optimal capital accumulation/income. As the time path of environment is determined conditional upon the autonomously chosen time path of income, one may say that in this case there is unidirectional causality from income to environment. Finally, consider the case where $U_{CC}\neq 0$, but $U_{CE}=0$. In this case the rate of change of consumption will be independent of the rate of change of environmental stock. Hence this case may be regarded as one of unidirectional causality from environment to income.

The above discussion should help explain the relevance of econometric tests of causality like the GCT for examining the nature of causality between income and emission based on a set of time series/panel data on the relevant variables. Given the *stationary* time series of a pair of variables, say x_t and y_t, the GCT examines whether or not a kind of *statistical feedback* exists between the two time series. More specifically, y_t is said to *fail to Granger cause* x_t, if the forecast of x_t conditional upon $x_{t-1}, x_{t-2},..., y_{t-1}, y_{t-2},...$ is *no better than* the forecast of x_t conditional upon $x_{t-1}, x_{t-2},...$ alone. In other words, when previous realizations of *y affect* the current realization of *x*, but previous realizations of *x do not affect* the current realization of *y*[61], causality is said to be unidirectional from *y* to *x*. The cases of bi-directional causality and absence of causality can accordingly be defined. The GCT, thus, provides an econometric procedure for examining causality in this dynamic sense[62].

[60] For simplicity, we assume natural depreciation rate of capital and natural recovery rate of the environment are zero.

[61] That is, *y* is autonomous and *x* is determined conditional upon *y* in terms of our earlier discussion.

[62] See, the Appendix for a description of the methodology of GCT used for the present exercise.

The GCT is a regression-based technique. For testing the null hypothesis that *x does not cause y,* the following *autoregressive distributed lag* regression equation is estimated:

$$y_t = \beta_0 + \beta_1 y_{t-1} + ... + \beta_k y_{t-k} + \gamma_1 x_{t-1} + ... + \gamma_k x_{t-k} + \varepsilon_t, t = 1,2,...,T \quad (5.5)$$

Where ε_t, a white noise, is the random disturbance term and $\beta_0, \beta_1, ..., \gamma_1, \gamma_2, ...$ are the regression parameters. If for this regression model $H_0 : \gamma_1 = \gamma_2 = ... = \gamma_k = 0$ is not rejected, the null hypothesis that *x does not cause y* is not rejected[63].

To illustrate interpretation of results obtained by applying the GCT to a given income-emission data set, let x_t and y_t denote *growth rate* of emission and income, respectively[64]. Taking *k=1*, the *income to emission* causality will be tested on the basis of estimate of the regression equation[65]

$$y_t = \beta_o + \beta_1 y_{t-1} + \gamma_1 x_{t-1} + \varepsilon_t, t = 1,2,...,T \quad (5.6)$$

Now, suppose the estimated result rejects $H_0 : \gamma_1 = 0$, implying thereby that emission *causes* income. Further, suppose the corresponding test of the null hypothesis that income *causes* emission gets rejected. Combining the two results, one gets a *unidirectional causality* from emission to income. As we illustrate below, for satisfactory interpretation of the results obtained, the regression results will also have to be taken into account.

Suppose for a given data set the *estimated* GCT regression equations[66] are $\hat{y}_t = \hat{a}_1 + \hat{b}_1 y_{t-1} (o < \hat{b}_1 < 1)$ and $\hat{x}_t = \hat{a}_2 + \hat{b}_2 x_{t-1} + \hat{c}_2 y_{t-1} (0 < \hat{b}_2 < 1)$. The first equation suggests that the income growth rate is autoregressive (hence autonomous) and varies around a mean growth rate of $\hat{a}_1$. The second equation suggests that the growth rate of emission, in addition to its own autoregressive movement, is significantly affected by the growth rate of income. Here, thus, is the typical story of a growing economy where any shock in income growth rate will cause a corresponding shock in the growth rate of emission – an *income causing emission* growth pattern.

Let us consider another example. In this case, let the estimated GCT regression equation for income growth rate be $\hat{y}_t = \hat{a}_1 + \hat{c}_1 x_{t-1}$ and that for emission growth rate be statistically non-significant (i.e., $x_t = \varepsilon_t$, a white noise). The first equation suggests that the income growth rate is determined by the emission growth rate. However, the emission growth rate, being a white noise, fluctuates around the zero level (implying thereby that the economy

---

63 The GCT is based on the assumption that the concerned time series are stationary. If they are not, appropriate differencing of the original time series is made to obtain stationary series. Note also that the regression model is of the autoregressive distributed lag form, the order of which has to be decided empirically.

64 That is, $x_t = \log(emission_t) - \log(emission_{t-1})$ and $y_t = \log(income_t) - \log(income_{t-1})$. Note that when the original time series data of income and emission have time trend, the logarithmic first differencing will generate the stationary time series required for the application of the GCT.

65 Note however that the choice of the value of *k* is an empirical question, which is sorted out by using appropriate econometric criterion.

66 Here the *hat* sign denotes a statistically significant parameter.

concerned has reached a stable level of emission). Now, if $\hat{c}_1$ is positive, any shock decrease/increase in the emission growth rate will cause a downward/upward jump of the income growth rate from its *stationary* level $\hat{a}_1$ which the economy maintains. In this case, thus, *emission causes income*. More importantly, a (temporary) rise in income growth rate can only be achieved only by accelerating the growth rate of emission. If, on the other hand, $\hat{c}_1$ is negative, a shock increase in the emission growth rate will cause the income growth rate to fall below its stationary level – possibly indicating an extreme case of *reverse causality* due to some kind of a sustainability problem.

The examples discussed above are only two of numerous possibilities that may emerge when the GCT is used to detect causality pattern hidden in income-emission data. The discussion above, however, has been made with a lot of abstractions. In reality the nature of income-emission causality that an economy will experience may depend upon a number of structural features of the economy concerned. For example, the sectoral composition of GDP/GNP of an economy may be an important determinant. If manufacturing activities are more *emission intensive* than services activities, say, an economy relying mostly on manufacturing production for income growth may fail to contain emission growth and in course of development may even experience *reverse causality*, if emission growth constrains income growth further.

Another important determinant of the nature of the income-emission causality for an economy may be the *openness* of the economy under study. Consider, for example, a small closed economy having a limited reserve of fuel resources. A threat of exhaustion of domestic fuel reserve may lead the economy to constrain use of fuel and thereby limit income growth. In other words, in this case emission may cause income. However, if the economy opens up and participate in international trade, the fuel constraint may get lifted as the economy may choose to produce whatever income it likes using fuel imported from elsewhere. In other words, openness of the economy may reverse the direction of causality from emission to income to income to emission. To put it differently, economies with *high openness* may display a tendency towards income to emission causality, while those less open, being constrained by their own natural resource endowment, may show emission to income causality.

A third major determinant of the nature of income-emission causality for an economy may be the price of fuel itself. As is well known, a drastic fuel price hike may make fuel an extremely scarce resource (particularly, for economies not having much of domestic reserve of fuel resource) and hence may reverse the nature of income-emission causality that prevailed prior to the fuel price hike. In fact, available evidences suggest significant changes in the rate of use of fuel by individual economies after the oil price shock of the 1970s.

5.3. THE DATA SET

In the present exercise we have used annual per capita real GDP *(PCGDP)* and annual per capita CO_2 *(PCCO2)* emission as the measure of the income and the emission variable, respectively[67]. The basic country-level time series data *PCGDP* (expressed in 1985 international prices, i.e., PPP dollars) for the period 1950-1992 were taken from the RGDPCH series of the Penn World Table (Mark 5.6) available at the web site http://www.nber.org/pwt5.6. This data set known as PWT5.6 is a revised and updated version of the preceding (Mark 5) version of the Penn World Table. The corresponding country-level annual time series data on *PCCO2* (expressed in metric tons) for the period 1950–1996 were obtained from the Tables of National CO_2 Emissions prepared by Carbon Dioxide Information Analysis Center (CDIAC), Environmental Science Division, Oak Ridge National Laboratory (ORNL) of the USA[68]. This data set happens to be the only available global data set on CO_2 emissions[69]. Combining both data sets together, we could compile a panel data set for 88 countries covering the period 1960–1990. The list of countries from different continents covered in the present exercise is presented in Table A.1 in Appendix A.

5.4. The Results

Using the cross country panel data set on income and emission described above and the methodology of GCT explained in Appendix B, we examined the income-emission causality patterns for country groups formed out of the 88 countries covered in the data set. To be precise, the basic exercise has been done separately for 12 country groups covering countries in different continents[70] (See, Table A.1 in Appendix A). Apart from this basic exercise, we also did some specific exercises to examine whether the nature of income-emission causality for an economy got affected by (i) the sectoral composition of income generated, (ii) the degree of openness of the economy and (iii) the fuel price regime. It may be noted that the present exercise was exploratory in nature and, therefore, we did not set *a priori* any null hypotheses to be checked and refuted empirically. In what follows, we present and explain the empirical results obtained by us.

5.4.1. Basic Results

Tables 5.1 and 5.2 present the basic results of application of GCT to the country group/continent-wise aggregate time series data and panel data, respectively. To be specific, the F-values reported in these Tables are those computed for testing the null hypothesis of absence of causality (for details, See Appendix B). For each country group/continent two F

[67] Per capita CO_2 emission has been used in most of the earlier studies. See, e.g.,Holtz-Eakin and Selden (1995), Shafik (1994), Sengupta (1997), Cole et al., (1997), Moomaw and Unruh (1997), de Bruyn et al., (1998), Unruh and Moomaw (1998), Tucker (1995).

[68] CDIAC throws up year-wise country-level data on emission of CO_2 from burning of fossil fuels and manufacture of cement for most of the countries of the world. These data set is prepared from data on net *apparent* consumption of fossil fuels (based on the World Energy data set maintained by United Nations Statistical Division) and World cement manufacture (based on cement manufacturing data set maintained by the U.S. Bureau of Mines). Emissions are calculated using global average fuel chemistry and usage.

[69] These estimates do not include bunker fuel used in international transport because of the difficulty of apportioning this fuel among the countries benefiting from such transport activities. See the web site http://cdiac esd.ornl.gov/epubs/ndpo30/ndpo301.htm for details.

[70] It may be noted that countries falling into the same group are more or less in similar state of economic development.

values corresponding to Models I and II are reported. Of these, the F value corresponding to Model I relates to the test of income to emission causality and the other F value corresponding to Model II relates to the test of emission to income causality. The causality results of Tables 5.1 and 5.2 are presented in a summary form in Table 5.3.

Table 5.1. Results of Granger Causality Test based on aggregate time series data for Country Groups/Continents

Country Group/ Continent	Model	F-value
Africa	I	0.250
	II	4.288**
North America	I	1.945
	II	0.251
Central America	I	1.983
	II	6.769***
South America	I	0.224
	II	0.660
America	I	1.990
	II	0.240
Japan	I	0.129
	II	2.076
Asia (excl. Japan)	I	1.894
	II	0.163
Asia	I	2.709*
	II	0.210
East Europe	I	4.416**
	II	1.233
Western Europe	I	0.877
	II	0.831
Europe	I	1.492
	II	0.708
Oceania	I	0.472
	II	5.111**
World	I	4.555**
	II	0.433

Note: In this and all other Tables that follow, *, ** and *** against the reported F-values denote significance at 10, 5 and 1 per cent level, respectively.

Qualitatively, the results in Table 5.1 based on country group/continent specific aggregate time series data are somewhat different from those in Table 5.2 based on the corresponding panel data sets. To be precise, Table 5.2 shows a much larger number of cases of significant causality – a result which is only to be expected, given the fact that a set of panel data contains much more information than the corresponding aggregate time series data set. As Tables 5.2 and 5.3 show[71], for North America, Eastern Europe and Western Europe

[71] It may be noted that in the cases of panel data set the GCT was performed using two methods of estimation, the OLS and the *within* regression method. The OLS and the *within* method were found to give very similar results.

the causality is unidirectional from emission to income. For Central and South America, America as a whole and Oceania, on the other hand, the causality is unidirectional from income to emission. For Africa, Asia excluding Japan, Asia as a whole, Europe and the whole World, significant bi-directional income-emission causality is observed[72]. Finally, income-emission causality is found to be absent for Japan.

Table 5.2. Results of Granger Causality Test based on OLS and *within* regression method applied to Panel data for Country Groups/Continents

Country Group/Continent	Model	F -value	
		OLS	*Within*
Africa	I	3.959**	3.624**
	II	5.598***	4.712***
North America	I	3.088*	3.407*
	II	0.728	0.726
Central America	I	1.638	1.918
	II	17.810***	18.124***
South America	I	1.113	0.959
	II	7.065***	7.164***
America	I	2.196	2.546*
	II	31.360***	32.913***
Japan	I	0.129	-
	II	2.076	-
Asia(excl. Japan)	I	3.001*	2.754*
	II	8.962***	6.789***
Asia	I	3.244**	3.113**
	II	9.970***	7.965***
East Europe	I	6.075***	6.005***
	II	2.059	2.803*
Western Europe	I	5.928***	5.298***
	II	2.057	1.398
Europe	I	11.410***	10.474***
	II	3.342**	3.072**
Oceania	I	1.383	0.703
	II	4.380**	5.824***
World	I	10.549***	11.027***
	II	32.272***	29.469***

Note: 1. As the question of a panel data set for Japan does not arise, here the results of Table 1 are reproduced.

2. Japan being a single country, the question of *within* estimation for Japan does not arise.

[72] To further probe the bi-directional causality results, we examined the nature of causality for individual countries of Asia and Africa. Among the countries of Asia income to emission causality was found for China, Iran, Philippines and Qatar, while causality in the reverse direction was observed for Hong Kong, Syria and Thailand. Among the African countries, income to emission causality was observed for Nigeria, Congo and Madagascar, while causality in the opposite direction was found for Gambia, Guinea-Bissau and Senegal. For the remaining countries of Asia and Africa no significant causality was detected.

Table 5.3. Summary Results of the Granger Causality Test

Country Group/ Continent	Aggregate data	Panel data (OLS)	Panel data (*within*)
Africa	income $\Rightarrow$ emission	income $\Leftrightarrow$ emission	income $\Leftrightarrow$ emission
North America	-	emission $\Rightarrow$ income	emission $\Rightarrow$ income
Central America	income $\Rightarrow$ emission	income $\Rightarrow$ emission	income $\Rightarrow$ emission
South America	-	income $\Rightarrow$ emission	income $\Rightarrow$ emission
America	-	income $\Rightarrow$ emission	income $\Leftrightarrow$ emission
Japan Ψ	income $\Rightarrow$ emission	-	-
Asia(excl. Japan)	-	income $\Leftrightarrow$ emission	income $\Leftrightarrow$ emission
Asia	emission $\Rightarrow$ income	income $\Leftrightarrow$ emission	income $\Leftrightarrow$ emission
East Europe	emission $\Rightarrow$ income	emission $\Rightarrow$ income	emission $\Rightarrow$ income
Western Europe	-	emission $\Rightarrow$ income	emission $\Rightarrow$ income
Europe	-	income $\Leftrightarrow$ emission	income $\Leftrightarrow$ emission
Oceania	income $\Rightarrow$ emission	income $\Rightarrow$ emission	income $\Rightarrow$ emission
World	emission $\Rightarrow$ income	income $\Leftrightarrow$ emission	income $\Leftrightarrow$ emission

Note: Blanks indicate cases of absence of any Granger causality.

5.4.2. Explanation of the Basic Results

In Table 5.4 the country group/continent-specific OLS estimates of the pair of GCT regression equations (i.e., equations (B.1) and (B.2) of Appendix B) based on panel data have been reproduced. We shall now attempt to use these regression results for explanation of the observed causality results.

As mentioned in Appendix B, the dependent variables of equation (B.1) and (B.2) of Appendix B are r_t and r_t^* (measuring growth rate of income and emission, respectively). So, in general, we may write equation (B.1) as $r_t = a + br_{t-1} + cr_{t-1}^* + \varepsilon_{1t}$ and equation (B.2) as $r_t^* = \alpha + \beta r_{t-1} + \gamma r_{t-1}^* + \varepsilon_{2t}$, where ε_{1t} and ε_{2t} are white noise error terms with zero expectations. Now, for a specific country group these equations take specific form depending on the statistical significance of the individual parameters of the above pair of equations. We discuss these cases below.

Consider first the cases of Africa and Asia for which all the estimated parameters are significant. Thus, r_t and r_t^* are interdependent, each being affected by the one period lagged values of both. In other words, for these country groups change in the income growth rate affects the growth rate of emission and vice versa. Hence, an observed bi-directional causality in these cases.

Next, let us consider the cases of North America and Western Europe. We have for North America $r_t = a + cr_{t-1}^* + \varepsilon_{1t}$ and $r_t^* = \varepsilon_{2t}$, and for Western Europe $r_t = a + br_{t-1} + cr_{t-1}^* + \varepsilon_{1t}$ and $r_t^* = \varepsilon_{2t}$. Thus, in both the cases the rate of growth of emission has reached a stage of *stationarity*, fluctuating randomly around the zero level. Now, for North America since r_t is a linear function of r_{t-1}^* (which is stationary), it is also stationary around a positive mean level *a* (estimated to be of the order of 0.01) in view of this linear relation. This implies that any shock in r^*_{t-1} will cause a corresponding shock in r_t. More importantly, if the emission rate is suddenly reduced, there will be a corresponding reduction in the income growth rate. However, since the income growth rate is not governed by an autoregressive effect, there will not be any persistent effect of the drop in the emission rate on the income growth rate. Hence, we have a very specific kind of emission to income *reverse* causality for North America.

Table 5.4. Country group/Continent-specific OLS estimates of eqs. (B.1) and (B.2) based on panel data

Country group/ Continent	Regression Equation for							
	r_t				r^*_t			
	Constant	r_{t-1}	r^*_{t-1}	$\overline{R^2}$	Constant	r_{t-1}	r^*_{t-1}	$\overline{R^2}$
Africa	0.0044	0.0823	0.0208	0.0128	0.0158	0.3273	-0.2818	0.0689
	(3.60)	(2.25)	(2.36)		(2.98)	(2.05)	(-7.32)	
North America	0.0094	0.0018	0.2449	0.0741	-0.0002	0.2434	0.1978	0.0619
	(4.44)	(0.01)	(2.14)		(-0.06)	(1.08)	(1.22)	
Central America	0.0035	0.1749	0.0195	0.0336	0.0103	0.8700	-0.3827	0.1646
	(2.38)	(2.93)	(1.08)		(2.25)	(4.70)	(-6.79)	
South America	0.0032	0.2687	0.0481	0.0885	0.0056	0.3633	-0.0155	0.0348
	(2.36)	(4.38)	(1.34)		(2.36)	(3.31)	(-0.24)	
America	0.0038	0.2256	0.0239	0.0597	0.0075	0.6865	-0.3146	0.1201
	(4.14)	(5.62)	(1.58)		(3.14)	(6.55)	(-7.99)	
Japan	0.0133	0.2599	0.1656	0.2245	-0.0094	1.1796	-0.0069	0.3864
	(2.54)	(0.86)	(0.97)		(-1.13)	(2.45)	(-0.02)	
Asia(excluding	.0128	0.0985	0.0578	0.0296	0.0174	0.2983	-0.1109	0.0169
Japan)	(7.59)	(1.88)	(2.60)		(4.33)	(2.38)	(-2.09)	
Asia	0.0130	0.1072	0.0586	0.0330	0.0166	0.3121	-0.1064	0.0178
	(8.16)	(2.12)	(2.72)		(4.39)	(2.60)	(-2.08)	
Eastern Europe	0.0096	0.1287	0.1400	0.0763	0.0107	0.1536	0.1162	0.0152
	(5.57)	(1.60)	(3.09)		(3.60)	(1.10)	(1.48)	
Western Europe	0.0096	0.2088	0.0582	0.0849	0.0034	0.2139	0.0633	0.0101
	(10.57)	(4.27)	(3.10)		(1.40)	(1.62)	(1.25)	
Europe	0.0098	0.1835	0.0738	0.0812	0.0055	0.1928	0.0833	0.0148
	(12.32)	(4.39)	(4.23)		(2.84)	(1.89)	(1.96)	
Oceania	0.0053	0.1716	-0.0244	0.0081	0.0103	0.8794	-0.2054	0.0717
	(2.93)	(1.68)	(-0.75)		(1.95)	(2.96)	(-2.16)	
World	0.0071	0.1564	0.0266	0.0361	0.0114	0.4209	-0.2501	0.0616
	(13.19)	(7.86)	(4.51)		(6.00)	(6.01)	(-12.03)	

Note: Figures in parentheses are the t-ratios.

For Western Europe, on the other hand, r_t, having a partial autoregressive effect in addition, is affected by both r_{t-1}^* and r_{t-1}. Thus, in this case the emission to income causality is supplemented by an additional autoregressive effect of income growth. This means that a sudden drop in the emission rate will cause not only a corresponding immediate negative shock in the income growth rate, the effect will linger due to the significant autoregressive element that governs the income growth rate. Clearly, thus, the emission to income *reverse* causality for Western Europe is qualitatively different from the nature of causality for North America that we have just noted.

For Eastern Europe the nature of the result is qualitatively somewhat different from those of North America and Western Europe. In this case we have $r_t = a + cr_{t-1}^* + \varepsilon_{1t}$ and $r_t^* = \alpha + \varepsilon_{2t}$ ($\alpha > 0$). Thus, here the growth rate of emission is stationary around a non-zero mean. Growth rate of income, being dependent on the growth rate of emission, is also stationary around a constant mean level. But, any shock in emission growth rate r_t^* will cause a fluctuation in the income growth rate. Hence, in this case also there is causality from emission to income. However, the case of Eastern Europe is qualitatively different from those of North America and Western Europe in the sense that while in the former case emission is observed to have a positive trend growth, in the latter cases this trend growth rate of emission has reached the zero level.

For Central America, we have $r_t = a + br_{t-1} + \varepsilon_{1t}$ and $r_t^* = \alpha + \beta r_{t-1} + \gamma r_{t-1}^* + \varepsilon_{2t}$. Here, r_t, following a first order autoregressive process, is clearly autonomous. On the other hand, r_t^* significantly depends upon both r_{t-1} and its own past value. Thus, we have a case of income to emission causality. The story for South America is very similar to this. But in that case γ (i.e., the autoregressive parameter of emission growth rate) is zero. Hence, the income to emission causality is stronger for South America.

Japan's case is somewhat unique. For this country we have $r_t = a + \varepsilon_{1t}$ ($a > 0$), which means a stationary income growth rate around a positive mean level. On the other hand, $r_t^* = \beta r_{t-1} + \varepsilon_{2t}$. These together mean a stationary emission around a positive mean level as well. However, any shock in income growth rate will result in a corresponding shock in the emission growth rate and in that sense there is an income to emission causality for Japan.

Finally, for Oceania we have $r_t = a + \varepsilon_{1t}$ ($a > 0$) and $r_t^* = \beta r_{t-1} - \gamma r_{t-1}^* + \varepsilon_{2t}, \beta, \gamma > 0$. Here the income growth rate is stationary around a positive mean level, while emission growth rate depends both on the income growth rate and the emission growth rate of the previous period. Thus, any shock in the income growth rate will get transmitted to the emission growth rate, thus resulting in an income to emission causality.

Let us next try to see the implications of the regression results of Table 5.4 for growth of emission in the long run equilibrium sense. Consider a given income growth rate r, say. From equation (B.2) we have $r_t^* = \alpha + \beta r + \gamma r_{t-1}^*$. The long run (dynamic) equilibrium emission growth rate is then $r_e^* = \dfrac{\alpha + \beta r}{(1-\gamma)}$. Since for North America, Eastern Europe, and Western Europe the estimated β and γ coefficients are not statistically significant, for these country groups $r_e^* = \alpha$ for these country groups. Further, while for Eastern Europe the estimated α is statistically significant, for the other two country groups it is not. Thus, the long run (dynamic) equilibrium emission growth rate would be zero for North America and Western Europe and a meagre 1 per cent for Eastern Europe. Incidentally, these are the country groups that showed significant unidirectional causality from emission to income.

Table 5.5. Country group-specific long run equilibrium growth rate of emission corresponding to a 10 per cent growth rate of income

Country group/continent	α	β	γ	r_e^* %)	r^{**}(%)
Africa	0.0158	0.3273	-0.2818	3.79	1.66
North America	-0.0002	0.2434	0.1978	3.01	-0.04
Central America	0.0103	0.8700	-0.3827	7.04	2.01
South America	0.0056	0.3633	-0.0155	4.13	0.86
America	0.0075	0.6865	-0.3146	5.79	1.19
Japan	-0.0094	1.1796	-0.0069	10.78	5.44
Asia (excl. Japan)	0.0174	0.2983	-0.1109	4.25	2.14
Asia	0.0166	0.3121	-0.1064	4.32	2.09
East Europe	0.0107	0.1536	0.1162	2.95	1.47
Western Europe	0.0034	0.2139	0.0633	2.65	0.47
Europe	0.0055	0.1928	0.0833	2.70	0.76
Oceania	0.0103	0.8794	-0.2054	8.15	3.16
World	0.0114	0.4209	-0.2501	4.28	1.37

It may be noted from Table 5.4 that for all the other country groups the estimated α, β and γ coefficients are statistically significant. Further, α and β are positive while γ is negative[73]. The negative value of γ suggests an oscillatory movement of the emission growth rate r_t^*, given the income growth rate. This is possibly due to the fact that fuel/energy input is mostly scarce and expensive for these country groups and therefore if fuel is over-consumed in one period, there will be an attempt to restrain its use in the next period. The long run (dynamic) equilibrium growth implication of this result is also quite interesting. Here, we have r_e^* greater than, equal to, or less than r according as r is less than, equal to, or greater than $r^{**} = \dfrac{\alpha}{(1-\gamma-\beta)}$. Thus, for countries/country groups for which income growth rate r exceeds the corresponding long run equilibrium emission growth rate r^{**}, an increasing and concave emission-income relationship is implied. On the other hand, for countries/country

[73] These results are econometrically stable in the sense that similar types of estimates were obtained for all the twenty one year sub-sample data sets covering the periods 1963-1983, 1964-1984, 1965-1985 etc.

groups with r less than the corresponding r^{**}, the long run equilibrium emission-income relationship would be increasing, but convex in income.

It should be pointed out here that the result for Japan does not fall into line with the other results. In this case the estimate of β is greater than 1 and statistically significant and the estimates of α and γ are non-significant. This means $r_e^* = \beta r$ and $r^{**} = 0$. Now, $\beta>1$ implies $r_e^* > r$ for all $r>0$. In other words, the long run equilibrium emission-income relationship is increasing and convex in income in this case – a result qualitatively different from what we have got for other country groups of the developed world.

Table 5.5 presents country group-specific estimates of the parameters α, β, γ of equation (B.2) and the corresponding r_e^* (based on the assumption that the income growth rate is r = 10 per cent) and r^{**} values[74]. By and large, these results show that the country groups of the developing world have both higher r_e^* and r^{**} values. This possibly reflects the relative inefficiency of these country groups *vis a vis* their developed counterparts in respect of fuel consumption/ emission.

5.4.3. Some Specific Results

As mentioned earlier, we tried to examine if the nature of income-emission causality might get affected by (1) the share of manufacturing in the GDP, (2) the degree of openness and (3) the level of oil price faced by an economy. One may argue that, manufacturing being more emission-intensive activity compared to, say, services, an economy having a large share of manufacturing in the GDP and continuing to depend on expansion of the manufacturing activities for income growth will have a stronger tendency towards to experience emission to income reverse causality. Degree of openness of an economy may also influence the nature of income-emission causality. To be specific, a highly open economy, because of its easy access to fuel through international trade, may not face the fuel supply constraint and hence continue to have the income to emission causality as income grows and thus avoid the reverse causality problem. Finally, *a priori*, a regime of high oil price, by forcing to economize the use of oil, may induce an emission to income causality in the economies all over. In other words, one might expect that the direction of causality would be from income to emission during periods of easy oil price and this would reverse during the high oil price periods.

To examine the effect of share of manufacturing in GDP on income-emission causality, we selected a set of 28 countries (out of the 88 countries in our data set) having 20 per cent or more share of manufacturing in GDP (as of the year 1986, see, World Development Report 1988). We then grouped these 28 countries by continents they belong to and formed four

[74] Here we have ignored whether the individual parameter estimates are statistically significant or not. If we had not done so and assigned zero values to the non-significant parameters, then, as already mentioned, r_e^* and r^{**} would be zero for North America and Western Europe. For Eastern Europe both r_e^* and r^{**} would be around 1 per cent. For other country groups the values of r_e^* and r^{**} would remain the same as reported in Table 4.5, as for them all the parameter estimates are statistically significant.

continent-specific country groups. The number of countries falling in Africa, America (covering North, Central and South America), Asia and Europe was 3, 8, 7 and 10, respectively. Using panel data, we examined the nature of income-emission causality for each of these four country groups. While for Asia and America significant income to emission causality was observed, a significant emission to income causality was observed for Europe. For Africa, however, no significant causality could be found. These results thus offered a mixed set of evidences so far as the effect of sectoral share of income on income-emission causality is concerned.

Next, we examined the effect of the degree of openness of an economy on the income-emission causality pattern. To do so, we classified the 88 countries of our data set into highly open, moderately open and narrowly open groups using the measure of openness given in the Penn World Table for the year 1985,[75]. Of the 15 highly open economies, income to emission causality was found for 2, emission to income causality was found for 1 and for the remaining 12 economies significant causality was found to be absent. On the other hand, of the 31 narrowly open economies, 11 showed income to emission causality, 3 showed emission to income causality, 2 showed bi-directional causality and for the remaining 15 causality was absent. This result thus, by and large, does not support the view that highly open economies will tend to have income to emission causality and narrowly open economies will tend to have emission to income causality.

Finally, we examined the effect of oil price regimes on the income-emission causality pattern. For doing this we repeated the GCT exercise for the three sub-periods of the original data set corresponding to the low oil price regime (the period 1960–73), the high oil price regime (the period 1974-79) and the post oil crisis period of declining oil price (the period 1980-90). The following results were obtained. For the low price regime, significant income to emission causality was observed for Africa, Central and South America, Western Europe and the World as a whole. Only for Europe as a whole significant emission to income causality was found. For the high price regime, significant income to emission causality was observed for Western and Eastern Europe, America as a whole and the World as a whole, causality being observed to be non-significant for all other country groups. Finally, for the post oil crisis period, income to emission causality was found for Central America, America as a whole and the World as a whole and emission to income causality was observed for both Asia and Asia excluding Japan. These results, on the whole, tend to reject the supposition that changes in the oil price regime may significantly reverse the pattern of income-emission causality. However, one should not attach much weight to these results because of the fact that these are based on data sets, which are possibly not adequate for application of a time series econometric tool like the GCT.

5.5. CONCLUSION

The results of income-emission causality study based on the GCT presented in this chapter may be summarized as follows: it is observed that for individual country groups well-

[75] The openness measure was defined to be (exports + imports) as ratio to GDP at current international prices. Countries with value of this ratio 100 or more, between 50 and 100 and less than 50 were classified as highly, moderately and narrowly open, respectively.

defined and distinctive patterns of causality prevail. Thus, for the developed country groups of North America and Western Europe (and for that matter, Eastern Europe also) the causality seems to run from emission to income. For the developing country groups of Central and South America and Oceania, on the other hand, the causality is found to run in the opposite direction from income to emission, and for Asia and Africa the causality turned out to be bi-directional. Japan, however, showed income to emission causality – a result that does not seem to match those observed for other countries of the developed world. On further probe, the bi-directional causality for Asia and Africa appeared to be due to the pooling of countries (with heterogeneous causality patterns) rather than being a phenomenon in itself. Our interpretation of the observed causality patterns made it clear how shocks in the rate of growth of income or emission might affect each other, depending on the prevailing nature of causality.

Let us briefly discuss the possible policy implications of the results presented in this chapter. Our results indicate that the countries of North America and Western Europe are in a stage of reverse causality in the sense that if the emission rate is suddenly reduced, there will be a corresponding reduction in the income growth rate. Moreover, since the growth rate of emission in these countries has reached a stationary zero level, any attempt to raise the income growth rate must require a corresponding growth in emission by these countries. On the other hand, if the emission rate is suddenly reduced, there will be a corresponding reduction in the income growth rate. The level of CO_2 emission of these countries is already very high. For example, the USA, which has the highest level of CO_2 emission (approximately 1490 million metric tons in 1997)) alone accounts for 23.58% and North America and Western Europe together account for 41.72% of the annual global emission (see Oak Ridge National Laboratory, CDIAC, 2000). This implies that if these countries wanted to have a marginal income increase by effecting a slight increase in the emission rate, that would mean quite large additional emission at the global level. On the other hand, as our results suggest, for the developing world causality between income and CO_2 emission is either absent or of the type income to emission. Further, available data suggest that the countries in this group are mostly small polluters (except China and India, which individually account for 14.47% and 4.43% of annual global emission of CO_2, respectively). Considering the fact that CO_2 emission generates externality at the global level, which is known to affect mankind universally, our results suggest two alternative policy conclusions. First, if the countries of the developed world want to maintain their current level of affluence or to raise it, they should seriously try to shift from fossil fuel to less polluting alternative so that the pollution due to CO_2 emission at the global level may be arrested. The alternative should be to evolve a system of tradable permits[76] at the international level, under which the countries of the developed world might buy permits for CO_2 emission from those of the developing world against compensatory payments. Such a policy, however, is unlikely to take shape easily in the absence of a global pollution monitoring and control agency. Even if implementation is possible, it should face tremendous objection from the developing countries, as they may not agree to sacrifice their freedom to develop and industrialize in exchange of dole receivable by sale of the tradable permits.

Let us next enumerate the limitations of the present study. First, a comprehensive analysis of income-emission relationship would necessarily call for examination of the effects

[76] See, Jensen and Rasmussen (2000).

of such determinants of fuel use as the sectoral composition of income, the openness of the economy and the price of fuel, among other things. In the present study, we tried to examine the effects of these possible determinants of income-emission relation only tangentially. One should take into account the above-mentioned factors while analysing the income-emission relationship using a much broader framework of study. Here we have abstracted from such an analysis, as our objective was essentially to see to what extent an econometric tool like the GCT could be utilized to draw useful conclusions about emission and environment from an emission-income data set. We hope to have a follow up study looking in to this aspect of the problem.

What would be the implications of the results of the present study for the current concern about global environmental changes, particularly those caused primarily by CO_2 emission? Given the rather long life of CO_2 molecules (about 100 years), a cleaner global environment and an arrest of the phenomenon of global warming would call for a check on the rate (of growth) of CO_2 emission at the global level. As the global rate of (growth of) CO_2 emission is an aggregate of the corresponding country-wise (or for that matter country group wise) rates, any policy formulation for the control of global CO_2 emission must pay attention to the country (group)-specific emission rates and their changes over time. Now, given that one or the other of the possible types of income-emission causality hold for individual country groups, any meaningful policy discussion for control of global CO_2 emission should require a careful examination of the cross-country distributional patterns of global income and the corresponding aggregate emission and their changes over time, keeping in mind the nature of causality that is operative in individual cases.

The notion of causality between income growth and pollution that underlies the *EKC* hypothesis is essentially a long run concept. Thus, further probe into the issue of causality using comprehensive econometric tools for exploring presence of any long run equilibrium relationship among income and pollution, viz., the cointegration analysis, may help to verify conclusions about causality that we have reached so far. In next Chapter, we shall be concerned with cointegration analysis. Thus, chapters 5 and 6, taken together will provide a more complete picture of causality.

Chapter 6

INCOME AND EMISSION: A PANEL DATA BASED COINTEGRATION ANALYSIS*

6.1. INTRODUCTION

In Chapter 5 we examined the nature of causality between CO_2 emission and income using the Granger causality test *(GCT)*. The *GCT* has been used in some empirical studies on *EKC* and related issues[77]. This technique alone, however, can detect presence and direction of causality for a pair of variables only in a limited sense (viz., in respect of their short run temporal movements). The notion of causality between income growth and pollution that underlies the *EKC* hypothesis, on the other hand, is essentially a longer run concept[78]. Thus, further probe into the issue of causality using comprehensive econometric tools for exploring presence of any long run equilibrium relationship among income and pollution, viz., the cointegration analysis, may help verify conclusions about causality that have been reached so far[79].

In this chapter, the results of an analysis of the relationship between per capita *GDP (PCGDP)* and per capita CO_2 emission *(PCCO2)* obtained by using non-stationary panel data techniques. The cross-country panel data set on these variables described in section 4.3 of the previous chapter. For convenience of exposition, henceforth these variables will be called *income* and *emission*, respectively. In this analysis, first the *panel data unit root test* of Im, Pesaran and Shin (2003) (henceforth referred to as IPS) has been performed to examine whether the observed country-specific time series data on income and emission possessed a stochastic trend or not. Next, on finding evidences of presence of such trend in the data set,

* This chapter is based on Dinda and Coondoo (2006).

[77] See, e.g., Cheng (1996), Cheng and Lai (1997) and Yang (2000).

[78] See, Chapter 5 for a discussion on this issue.

[79] There are interesting applications of time series econometric tools like vector autoregressive model (VAR) and cointegration analysis on environment-related data. See, e.g., Stern (1993, 2000) for studies on causal relationship between GDP and energy use for the USA for the period 1947-1990 based on *GCT* in a VAR set up, single equation static cointegration analysis and multivariate dynamic cointegration analysis. See also Cheng (1999) for an application of Johansen cointegration test to the data on energy consumption, economic growth, capital and labour for the Indian economy.

the Engle-Granger bivariate cointegration analysis[80] has been done to examine whether the pair of variables is cointegrated (i.e., whether they obey any long run equilibrium relationship between themselves). Finally, the *Error Correction Model (ECM)* has been estimated to explore the nature of dynamics implicit in the panel data set for those country-groups for which income-emission cointegration is obtained.

The chapter is organized as follows: section 6.2 explains the motivation for using cointegration analysis on the income-emission data in the present exercise; section 6.3 describes the data, presents and discusses the empirical results, section 6.4 interprets the results and section 6.5 draws some concluding observations. Finally, the methodology of unit root test, cointegration analysis and *ECM* estimation based on panel data that have actually been used in the present exercise are briefly explained in the Appendix B.

6.2. MOTIVATION

To help justify the use of cointegration analysis on the set of cross-country panel data on income and emission for examining the nature of causality that may exist between this pair of variables, let us reconsider the simple theoretical construct already developed in chapter 5 and rewrite equation (5.4) as

$$\alpha \frac{\dot{C}}{C} + \beta \frac{\dot{E}}{E} = \phi \tag{6.1}$$

where $\alpha = \frac{CU_{CC}}{U_C}$, $\beta = \frac{EU_{CE}}{U_C}$ and $\phi = (-\frac{f_K g_K}{g_K + \gamma f_K} + \rho)$, U_{CC}, U_{CE} being the second order partial derivatives of $U(.)$. It should be noted that the above condition suggests that optimal time path of C and E should generally be interdependent. This, thus, means a two-way causal relationship between income and environment, in general.

Let us next search for a long run equilibrium relationship between income (C) and environment (E), underlying the said optimization problem. To do so, consider the *steady state solution* where $\dot{E} = \dot{\mu} = 0$ i.e., the situation where the environmental stock reaches a stable level. Now, $\dot{E} = 0$ implies (from equation (5.3))

$$g((1-\theta)K) = \gamma f(\theta K, E) \tag{6.2}$$

i.e., the rate of environmental degradation due to production must equal the rate of environmental upgradation. Clearly, equation (6.2) defines a relationship between K and E – say,

[80] Johansen's method of cointegration analysis, which is more comprehensive, could not be used as we could not access the software required for application of this method to a panel data set.

$$h_1(K,E) = 0, \tag{6.3}$$

for given θ. Next, let at the steady state $\dot{K} = \sigma$, a constant. This implies (from eq. (5.2))

$$f(\theta K, E) - C = \sigma \Rightarrow h_2(K,E,C) = 0, \tag{6.4}$$

for given θ. Combining equations (6.3) and (6.4), we obtain what may be called a long run equilibrium relationship between *C* and *E*, say,

$$h_3(C,E) = 0, \text{ or equivalently, } E = h(C), \tag{6.5}$$

which may be recognized as the long run relationship between income (C) and environment (E). It may be noted that this long run environment-income relationship may or may not be an EKC. In this context, it should be mentioned that Brock and Taylor (2004a, 2004b) provide different possible scenario for the existence of EKC in their Green Solow model. In the Green Solow model (Brock and Taylor (2004a)) the EKC is a necessary by-product of convergence to a sustainable growth path. The resulting EKC may be humped shaped or strictly declining. The model outlined above may produce EKC in the transitional phase under specific conditions[81]. Our interest here is merely to examine whether or not the observed temporal movements of C and E obey an underlying stable relationship linking these two variables, following the spirit of cointegration analysis.

It should now be straightforward to use the above theoretical construct to rationalize cointegration analysis of a bivariate time series/panel data set on income and emission, as has been done in the present chapter. Let $\{C_t^*, E_t^*\}$ denote time series of observed consumption and environment variable, where $C_t^* = C_t + \varepsilon_{Ct}$ and $E_t^* = E_t + \varepsilon_{Et}$ - C_t, E_t being the corresponding (unobserved) optimal values and $\varepsilon_{Ct}, \varepsilon_{Et}$ being random disturbances. In case the observed data set is consistent with optimization, C_t^* and E_t^* should differ from the corresponding optimal values only by stationary random disturbances (i.e., ε_{Ct} and ε_{Et} should be stationary random variables). Also, C_t^* and E_t^*, being consistent with optimization, should be *cointegrated* as they must obey equation (5.5), but for stationary deviations.

Granger causality between *C* and *E*, which is essentially a short run notion, is often examined with the help of the *ECM* as a part of the cointegration analysis. When time series C_t^* and E_t^* are non-stationary and are integrated of order one (i.e., the corresponding time series of first differences are stationary) and the variables are cointegrated, they admit the *Granger representation*[82] and the *ECM* can be expressed as

$$\Delta C_t^* = \sum_{i=1}^{m} \beta_{Ci} \Delta C_{t-i}^* + \sum_{i=1}^{m} \gamma_{Ci} \Delta E_{t-i}^* - \eta_C (E_{t-1}^* - h(C_{t-1}^*)) + \nu_{Ct} \tag{6.6}$$

[81] See section 5 of Dinda (2005) for details.
[82] See Hamilton (1994) for details.

or, equivalently as

$$\Delta E_t^* = \sum_{i=1}^{m} \beta_{Ei} \Delta C_{t-i}^* + \sum_{i=1}^{m} \gamma_{Ei} \Delta E_{t-i}^* - \eta_E (E_{t-1}^* - h(C_{t-1}^*)) + \nu_{Et} \quad (6.7)$$

where ν_{Ct} and ν_{Et} are pure white noise random disturbances and $\beta_{Ci}, \beta_{Ei}, \gamma_{Ci}, \gamma_{Ei}, \eta_C$ and η_E are the parameters of the *ECM*. It may be noted that $(E_{t-1}^* - h(C_{t-1}^*))$, which is called the *error correction (EC)* term, is a measure of the extent by which the observed values in time *t-1* deviate from the long run equilibrium relationship. Since the variables are cointegrated, any such deviation at time *t-1* should induce changes in the values of the variables in the next time point in an attempt to force the variables back to the long run equilibrium relationship. The coefficients η_C and η_E of the error correction term in the two equations (which measure the rate of this adjustment process) are therefore called the *adjustment parameters* and are expected to be positive. The parameters γ_{Ci}'s in equation (6.6) and β_{Ei}'s in equation (6.7) determine the nature of causality between *C* and *E*. More specifically, if $\gamma_{Ci} \neq 0$ for at least one $i (i = 1, m)$ and $\beta_{Ei} = 0$ for all $i (i = 1, m)$, then *E* is said to *Granger cause* C. On the other hand, if $\gamma_{Ci} = 0$ for all $i (i = 1, m)$ and $\beta_{Ei} \neq 0$ for at least one $i (i = 1, m)$, then *C* is said to *Granger cause E*. In case $\gamma_{Ci} \neq 0$ and $\beta_{Ei} \neq 0$ for at least one $i (i = 1, m)$, the causality between *C* and *E* is defined to be bi-directional.

Finally, when $\gamma_{Ci} = 0$ and $\beta_{Ei} = 0$ for all $i (i = 1, m)$, Granger causality between *C* and *E* is said to be absent[83]. The absence of Granger Causality for cointegrated variables requires the additional condition that the speed of adjustment coefficient be equal to zero. In this set up, statistical significance of the estimated adjustment parameters η_C and η_E should help qualify further the nature of causality relationship between *C* and *E*. Thus, for example, if $H_0 : \beta_{Ei} = 0$ for $i = 1, m$, η_E=0 is not rejected and at the same time $H_0 : \gamma_{Ci} = 0$ for all $i (i = 1, m)$, η_C=0 is rejected, one should interpret such a result as corresponding to a situation in which the time path of *C* is autonomously determined and that of *E* being caused by *C*. Other possible results may be interpreted in a similar manner (see Glasure and Lee (1997) and Asafu-Adjaye (2000) for details).

[83] For the form of null hypotheses that have been tested to detect the nature of causality in the *ECM* set up, see Section A.3 of the Appendix.

6.3. THE RESULTS

The empirical exercise reported in this chapter has been done separately for each of these country-groups based on the bivariate panel data sets for the individual country-groups[84]. In this context, it may be mentioned that here the countries have been grouped essentially keeping in view their geographic contiguity. Since such a grouping of countries does not guarantee that all countries of a group have comparable levels of economic development (and the same nature of cointegration between *PCGDP* and *PCCO2*), the possibility that the country-group level cointegration results may be biased and distorted due to pooling of heterogeneous units into the same group cannot be ruled out, in principle. To verify whether the specific country-grouping that has been considered in the present exercise has significantly affected the results reported, the test of cointegration has been performed on a rearranged panel data set as well, which is based on an alternative set of country-groups formed using a rule that is *econometrically more* satisfactory. The results of the test of stationarity of variables and the subsequent test of cointegration for this alternative country-grouping are presented in Table 6.6[85].

Table 6.1. Results of Panel Unit Root Test: IPS $\bar{t}$ statistic by Country-group

Country Group	With Time Trend			Without Time Trend		
	t-bar for		Critical Value	t-bar for		Critical Value
	income	emission	(5% level)	income	emission	(5% level)
Africa	-0.289	-0.376	-2.45	2.469	0.664	-1.82
North America	-0.330	0.486	-2.94	0.296	-1.384	-2.30
Central America	2.109	1.025	-2.60	-0.038	-0.302	-1.99
South America	1.912	0.980	-2.60	1.210	0.949	-1.99
America	2.611	1.498	-2.47	0.880	0.019	-1.84
Japan	NA	NA		NA	NA	
Asia(excl. Japan)	-0.734	-0.250	-2.56	6.068	2.351	-1.94
Asia	-0.842	-0.307	-2.54	5.757	2.075	-1.92
East Europe	3.238	1.308	-2.74	-0.592	-2.123	-2.12
West Europe	-0.701	-0.605	-2.52	3.283	0.022	-1.89
Europe	1.093	0.167	-2.47	2.491	-1.090	-1.84
Oceania	-0.250	-0.488	-2.84	0.949	0.293	-2.21
World	1.306	0.402	-2.32	5.526	0.715	-1.68

Note: 1. Im *et al* (2003) provide Tables of critical values of Panel unit root test statistic for selected combinations of N and T values. The critical values shown in the present Table have been derived from the original Tables by interpolation wherever required. 2. NA denotes *not available*. For Japan, a single country, the panel unit root test was not applicable. Hence no result is shown against Japan.

---

[84] It may be mentioned that the states/regions covered by the erstwhile U. S. S. R. have been left out of this exercise because past data for these states/regions are not available. It should be noted that these states/regions/countries are more or less in a comparable state of economic development.

[85] We are grateful to an anonymous referee who pointed out this possibility and suggested the econometric classification rule that has been mentioned above.

Table 6.2. Results of cointegration Test: IPS $\bar{t}$ statistic by country-group

Country Group	Without time trend		Critical value	With time trend		Critical value
	income	emission		income	emission	
Africa	-0.880	-2.571***	-1.82	-2.643***	-4.18***	-2.45
North America	-0.608	-2.182	-2.30	-1.665	-0.567	-2.94
Central America	-2.015**	-2.263***	-1.99	0.905	-2.524*	-2.60
South America	-0.846	-1.091	-1.99	-1.384	-2.123	-2.60
America	-2.112**	-2.919***	-1.84	-0.825	-3.304***	-2.47
Japan	NA	NA		NA	NA	
Asia(excl. Japan)	3.428	-0.054	-1.94	-1.862	-1.543	-2.56
Asia	3.052	-0.398	-1.92	-1.879	-1.513	-2.54
East Europe	-2.089	-3.523***	-2.12	-2.237	-4.649***	-2.74
West Europe	0.572	-2.484***	-1.89	-3.088***	-3.935***	-2.52
Europe	-0.603	-3.958***	-1.84	-3.802***	-5.784***	-2.47
Oceania	-0.363	-0.978	-2.21	-0.922	-1.520	-2.84
World	-0.696	-5.203***	-1.68	-4.697***	-7.744***	-2.32

Note: *, ** and *** denote the significance level at 10%, 5% and 1%, respectively. Critical values shown correspond to the 5% level of significance. NA denotes "Not Applicable".

Table 6.1 presents the country-group-specific results of unit root test for logarithm of *PCGDP* and logarithm of *PCCO2* based on the *IPS* method. In each case the test has been done twice – viz., once assuming presence of a deterministic time trend in the data generating process and again assuming absence of such a trend. The results show that at 5 per cent level of significance the null hypothesis of unit root cannot be rejected in any of the cases, except for income for Eastern Europe when presence of a deterministic time trend in the data generating process is not assumed[86]. One may thus conclude that the country-group-specific time series of both the variables under consideration are by and large non-stationary. A repetition of the same test on the first-differenced data set results in rejection of the null hypothesis of unit root in all the cases. The results of unit root test thus indicate that the country-specific time series of both income and emission are integrated of order 1(i.e., they were *I(1)*, symbolically).

In the next step, we have examined whether or not for individual country-groups the null hypothesis that income and emission are *not cointegrated* may be rejected. As explained in the Appendix B, the bivariate *Engle-Granger* methodology of cointegration[87] and the *IPS* unit root test procedure has been used for this examination. The results of these tests are presented in Table 6.2. Following the *Engle-Granger* convention, for each country-group we have tested cointegration twice, viz., once treating income as the dependent variable and emission as the independent variable and again interchanging the dependent-independent status of these two variables. The entries under the column heading income (emission) are the

[86] In this case the test turned out to be marginally significant at the 5 per cent level in the *without time trend* case and was non-significant in the *with time trend* case. Such a result may be possible only if an increasing (decreasing) deterministic time trend gets canceled with a decreasing (increasing) stochastic time trend.

[87] In Engle and Granger's (1987) original definition, cointregation relates to a linear relationship between non-stationary variables. Holtz Eakin and Selden (1995) show an empirical evidence of such a linear relationship between per capita income and CO_2 emission. In the present exercise, we also find a significant relationship between income and emission.

computed *IPS t*-statistic values for the *cointegration unit root test* when income (emission) is taken as the dependent variable. Here also in each case the cointegration test[88] has been done twice – viz., once assuming presence of a deterministic time trend in the residuals of the cointegrating regression equation and again without making such an assumption. In Table 6.2 country-group-specific values of these four test statistics are presented.

Table 6.3. Estimated parameters of the *ECM* for country-groups for which cointegration hypothesis was not rejected

Country-group	Model	Estimated coefficient of the explanatory variable (Δ log)						
		income_1	income_2	income_3	emission_1	emission_2	emission_3	EC term
Africa	I (3)	0.10	0.10	-0.07	0.00	-0.00	-0.02	-0.09
		(2.59)	(2.56)	(-1.79)	(0.02)	(-0.25)	(-1.92)	(-4.77)
	II (3)	0.05	0.21	-0.18	-0.20	-0.08	-0.17	-0.26
		(0.31)	(1.31)	(-1.12)	(-4.59)	(-1.72)	(-4.12)	(-7.95)
Central	I(2)	0.192	0.019	-	0.004	0.016	-	-0.0906
America		(2.993)	(0.3)		(0.18)	(0.8)		(-2.62)
	II(2)	0.782	0.152	-	-0.4	-0.281	-	-0.186
		(4.11)	(0.81)		(-5.97)	(-4.7)		(-3.42)
	I(2)	0.229	-0.02	-	0.012	0.011	-	-0.059
America		(5.31)	(-0.45)		(0.7)	(0.68)		(-3.16)
	II(2)	0.666	0.191	-	-0.36	-0.238	-	-0.091
		(6.09)	(1.71)		(-8.14)	(-5.76)		(-3.28)
	I(3)	0.172	-0.056	0.205	0.052	-0.138	-0.018	-0.083
Eastern		(2.18)	(-0.72)	(2.7)	(1.09)	(-2.98)	(-0.37)	(-4.58)
Europe	II(2)	0.029	0.145	-	0.014	-0.018	-	-0.132
		(0.22)	(1.11)		(0.19)	(-0.24)		(-4.85)
	I (2)	0.24	-0.18	-	0.04	-0.04	-	-0.03
		(4.85)	(-3.62)		(1.97)	(-2.12)		(-3.33)
Western	II (2)	0.16	0.08	-	0.05	-0.07	-	-0.03
Europe		(1.12)	(0.59)		(0.98)	(-1.23)		(-1.74)
	I (3)	0.23	-0.13	0.08	0.04	-0.06	-0.02	-0.04
		(5.26)	(-2.92)	(1.81)	(2.32)	(-3.30)	(-0.76)	(-4.89)
	II (2)	0.12	0.11	-	0.07	-0.04	-	-0.07
Europe		(1.10)	(1.06)		(1.52)	(-0.97)		(-4.00)
	I (3)	0.12	0.03	-0.03	0.02	-0.00	-0.01	-0.04
		(5.45)	(1.55)	(-1.47)	(1.9)	(-0.12)	(-1.66)	(-5.33)
	II (3)	0.26	0.27	-0.02	-0.22	-0.09	-0.12	-0.17
World		(3.54)	(3.76)	(-0.32)	(-9.40)	(-4.06)	(-5.44)	(-11.12)

Note: 1. Figure in brackets in the "model" column indicates the optimum number of lagged variables used as regressors in the *ECM* as determined for the given data set. 2. For each country-group and model the first row of 3rd to 9th column gives the estimated coefficients. The corresponding figures in brackets in the next row of these columns are the corresponding t-ratios.

[88] That is, the unit root test of the residuals of the estimated long run relationship between *y* and *x*.

Table 6.2 may be summarized as follows: The results of cointegration appear to be sensitive to whether or not presence of a deterministic time trend in e_{it} 's (i.e., the regression residuals defined in relation (C.3) of the Appendix B) is assumed. When e_{it} 's are assumed not to contain a deterministic time trend, in most of the cases the result of cointegration is observed to depend upon whether income or emission is taken as the dependent variable. Exceptions are Central America, America as a whole and Eastern Europe. In all these cases the null hypothesis of no cointegration is rejected irrespective of whether income or emission is used as the dependent variable. In contrast, when presence of a deterministic time trend in e_{it} 's is assumed, the cointegration results obtained by treating income as the dependent variable are seen mostly to agree with the corresponding results obtained by treating emission as the dependent variable[89]. Thus, in this case, irrespective of whether emission or income is taken as the dependent variable, the null hypothesis of cointegration is not rejected for Africa, Western Europe, Europe and the World. In other words, for these country-groups time series of income and emission seem to obey a long run equilibrium relationship. For North America, South America, Asia, Asia excluding Japan and Oceania, on the other hand, the null hypothesis of cointegration is rejected. For the remaining country-groups (viz., Central America, America and Eastern Europe), the null hypothesis of cointegration is not rejected when emission is taken as the dependent variable, but it is rejected when income is taken as the dependent variable.

Next, using the country-group-specific panel data, we have estimated the alternative versions of the *ECM* - viz., equation (C.5) and (C.6) of the Appendix B, which henceforth we shall refer to as models I and II, respectively. This estimation has been done only for those country-groups for which the null hypothesis of cointegration is not rejected (viz., Africa, Central America, America as a whole, Eastern Europe, Western Europe, Europe as a whole and the World). In case of each of these country-groups, the *ECM* is estimated using three different econometric specifications of the panel data regression equation – viz., ordinary least squares (OLS), fixed effects (FE) model and random effects (RE) model[90]. As our results show, the FE model turns out to be the appropriate choice for almost all the country-groups. The country-group-specific estimates of the regression coefficients of the two versions of the *fixed effects ECM* (viz., models I and II) are presented in Table 6.3.

It may be noted that the estimated adjustment parameters (i.e., the coefficient of the *EC term*) in Table 6.3 are all statistically significant with the *expected* negative sign (in all cases except for Western Europe when emission is taken as the dependent variable). Since in all these cases income and emission are cointegrated, such a result is only to be expected. This is

---

[89] It is well known that in Engle-Granger methodology the result of the cointegration test may be sensitive to the choice of the dependent variable of the cointegration regression in case of *not large enough* samples. The power of the unit root test may also depend on whether or not a deterministic trend is present in the data generating process and has been incorporated in the regression model used to test unit root. When the regression model estimated for testing unit root contains a deterministic trend component and the test rejects the null hypothesis of presence of a unit root, that may be a sufficient indication of absence of a unit root (see, Enders (1995) pp. 254-258).

[90] OLS is known to be generally inefficient for panel data regression estimation. Choice between FE and RE depends upon whether or not the null hypothesis $H_0 : \alpha_i = \alpha$ for $i = 1,2,..., N$, is rejected, where α_i denotes the intercept for the *ith* unit. FE is chosen when H_0 is not rejected. For detail discussion see Baltagi (1999) and Hsiao (1986).

because of the following reason: as the pair of variables is cointegrated, over a long period of time they tend to move in unison always trying to be on the *long run equilibrium* relationship.

As is well known, the *ECM* tries to explain the observed short run variations of the dependable variable in terms of variations of the lagged value of the dependent variable and the other explanatory variable of the model. Following the explanation given in Section 5.2 and the Appendix B, the nature of *Granger causality* between the variables under study underlying the given data set may be examined by testing null hypotheses specifying relevant parametric restrictions on the estimated *ECM* (See Table 6.5a).

6.4. Interpretation of Results

In Table 6.3 the country-group/continent-specific FE estimates of the pair of *ECM* equations (i.e., equations (C.5) and (C.6) of Appendix B) based on panel data are presented. We shall now attempt to explain the results of Table 6.3 from the point of view of causality[91] due to short run fluctuations along with long run equilibrium relationship.

As is well known, the *ECM* describes the short-run dynamics of the variables of a system when the concerned variables violate the equilibrium relation(s) governing their long run movements.

The dependent variables of equation (C.5) and (C.6) of Appendix B are r_t and r_t^* measuring growth rate[92] of income and emission, respectively. So, in general, we may write equation (C.5) as $r_t = \sum_{i=1}^{T_{11}} \alpha_{1i} r_{t-i} + \sum_{j=1}^{T_{12}} \beta_{1j} r_{t-j}^* + \eta_y EC_{t-1} + u_{1t}$ and equation (C.6) as $r_t^* = \sum_{i=1}^{T_{21}} \alpha_{2i} r_{t-i} + \sum_{j=1}^{T_{22}} \beta_{2j} r_{t-j}^* + \eta_x EC_{t-1} + u_{2t}$, where *EC* denotes the error correction term, u_{1t} and u_{2t} are white noise error terms. As we have already seen, the estimated coefficient of the *EC term* in Table 6.3 are statistically significant with an *expected negative* sign (in all cases except for Western Europe for which significance level is low (viz., 10%), when emission is taken as the dependent variable). Now, for a specific country-group these equations take specific form depending on the statistical significance of the individual parameters of the above pair of equations. We discuss these cases below and also examine their implications for short run movement from the point of view of causality.

Consider first the case of Africa for which not all the estimated parameters are significant. We have $r_t = \alpha_1 r_{t-1} + \alpha_2 r_{t-2} - \eta_y EC_{t-1} + u_{1t}$, $\alpha_1, \alpha_2, \eta_y > 0$ and $r_t^* = -\beta_1 r_{t-1}^* - \beta_3 r_{t-3}^* - \eta_x EC_{t-1} + u_{2t}$; $\beta_1, \beta_3, \eta_x > 0$. Thus, r_t and r_t^* follow autoregressive processes and are autonomous in short run, although a statistically significant long run relationship exists between them.

---

[91] It should be noted that in our earlier study (i.e., Chapter 4) we found significant causality between income and emission using *GCT*. That result remains the same in this study when short run movements are considered, but differ when long run movements are also taken into account.

[92] $\Delta y_t = \Delta \ln(PCGDP) = r_t$ and $\Delta x_t = \Delta \ln(PCCO2) = r_t^*$.

For both Central America and America as a whole, we have $r_t = \alpha_1 r_{t-1} - \eta_y EC_{t-1} + u_{1t}$ and $r_t^* = \alpha_1 r_{t-1} - \beta_1 r_{t-1}^* - \beta_2 r_{t-2}^* - \eta_x EC_{t-1} + u_{2t}$; $\alpha_1, \beta_1, \beta_2, \eta_x > 0$. Here, r_t, which follows a first order autoregressive process, is clearly autonomous. On the other hand, r_t^* significantly depends upon both r_{t-1} and its own past values. Thus, we have a case of income to emission causality in the short run.

Next, let us consider the cases of Western Europe. We have $r_t = \alpha_1 r_{t-1} - \alpha_2 r_{t-2} + \beta_1 r_{t-1}^* - \beta_2 r_{t-2}^* - \eta_y EC_{t-1} + u_{1t}$; $\alpha_1, \alpha_2, \beta_1, \beta_3, \eta_y > 0$ and

$r_t^* = -\eta_x EC_{t-1} + u_{2t}$ (here the coefficient of EC term is significant at 10% level). These results suggest that the rate of growth of emission has reached a stage of *stationarity* maintaining a long run equilibrium relationship with the rate of growth of income, but in short run r_t significantly depends on both its own past value and r_{t-1}^*. This implies that any shock in r_{t-1}^* will cause a corresponding shock in r_t. Hence, we have a very specific kind of emission to income *reverse* causality for Western Europe.

Finally, we have $r_t = \alpha_1 r_{t-1} + \alpha_2 r_{t-3} - \beta_2 r_{t-2}^* - \eta_y EC_{t-1} + u_{1t}$; $\alpha_1, \alpha_2, \beta_2, \eta_y > 0$ for East Europe and $r_t = \alpha_1 r_{t-1} - \alpha_2 r_{t-2} + \beta_1 r_{t-1}^* - \beta_2 r_{t-2}^* - \eta_y EC_{t-1} + u_{1t}$; $\alpha_1, \alpha_2, \beta_1, \beta_2, \eta_y > 0$ for Europe as a whole; and $r_t^* = -\eta_x EC_{t-1} + u_{2t}$ for both. Thus, here the growth rate of emission is stationary with a long run equilibrium relationship. Growth rate of income, being dependent on the growth rate of emission, is also stationary, but any shock in emission growth rate r_t^* is likely to cause a fluctuation in the income growth rate. Hence, in these cases also there is a reverse causality from emission to income. However, in these cases the emission to income causality is supplemented by an additional autoregressive effect of income growth. This means that a sudden drop in the emission rate will cause not only a corresponding immediate negative shock in the income growth rate, the effect will linger due to the significant autoregressive element that governs the income growth rate.

Now, let us see the long run income-emission relationship (as given by the estimated cointegrating vector, viz., *(1, -b₀, -b₁)*) and also the speed of adjustment (η) for different country-groups. As is well known, the cointegrating vectors give the long run relationship between income and emission for individual country-groups. The cointegrating vectors[93] for Africa, Central America, America as a whole, Eastern Europe, Western Europe, Europe as a whole and the World as a whole are presented in Table 6.4.

---

[93] In Table 6.4, for every individual country-group a pair of cointegrating vectors has been reported. These have been obtained by changing the status of dependent and independent variables. Standard normalization process slightly differs in these cases because of the presence of country effects or some other fluctuations, although both the variable are cointegrated for individual country-groups.

Table 6.4. Country-group-specific estimated cointegrating relationships in the form

$$y_{it} = b_0 + b_1 x_{it} \text{ and } x_{it} = b_0' + b_1' y_{it}$$

Country-group	Dependent variable: y		Dependent variable: x	
	$-b_0$	$-b_1$	$-b_0'$	$-b_1'$
Africa	-7.88 (.0262)	-0.36 (.00927)	15.076 (.3214)	-1.80 (.0459)
Central America	-8.39 (.0231)	-0.46 (.0149)	14.13 (.4209)	-1.65 (.0535)
America	-8.53 (.0112)	-0.54 (.00779)	14.0 (.192)	-1.63 (.0237)
Eastern Europe	-8.33 (.0393)	-0.36 (.046)	5.59 (.7477)	-0.697 (.0884)
Western Europe	-8.71 (.0206)	-0.45 (.0214)	8.78 (.4539)	-1.05 (.0502)
Europe	-8.58 (.0205)	-0.48 (.022)	7.02 (.3483)	-0.86 (.0392)
World	-8.46 (.00898)	-0.55 (.00476)	12.96 (.1049)	-1.51 (.0131)

Note: Figures in parentheses are standard errors. All the estimated coefficients are statistically significant.

Table 6.5a. Computed F values for test of parametric restriction on the *ECM* relating to the *GCT*

Country-group	Model (lag)	OLS regression	FE regression	RE regression
Africa	I(3)	2.11	1.35	1.45
	II(3)	4.70**	1.07	1.37
Central America	I(2)	1.30	0.35	0.86
	II(2)	16.67**	9.07**	13.34**
America	I(2)	1.40	0.36	0.71
	II(2)	29.15**	21.50**	24.82**
Eastern Europe	I(3)	3.33*	3.80*	3.32*
	II(2)	2.52	0.68	1.75
Western Europe	I(2)	5.01**	4.62**	8.22**
	II(2)	1.73	0.96	4.76**
Europe	I(3)	6.15**	6.20**	6.05**
	II(2)	2.99	1.37	1.83
World	I (3)	4.50**	2.82**	3.47**
	II(3)	24.96**	9.53**	15.46**

Notes: (1). Models I and II relate to the *ECM* equations (C.5) and (C.6) of the Appendix. (2). Figures in parentheses give the order of the ECM regression equation in terms of the maximum order of lag of variables appearing as regressors. (3). For model I and II the computed F value relates to the null hypothesis $\beta_{1j} = 0$ for all j and $\eta_{yx} = 0$ and $\alpha_{2j} = 0$ for all j and $\eta_{xy} = 0$, respectively. (4). F- values marked by * and ** are significant at 5 and 1 per cent level, respectively.

Table 6.5b. Estimated values of Error Correction term for different models in panel data

Country-group	Model	Pooled(OLS)	Fixed Effect	Random Effect
Africa	I	-0.019** (-2.27)	-0.087*** (-4.77)	-0.027*** (-2.7)
	II	-0.054*** (-3.24)	-0.263*** (-7.95)	-0.125*** (-5.36)
Central America	I	-0.012 (-0.92)	-0.091*** (-2.62)	-0.015 (-1.06)
	II	-0.022 (-1.08)	-0.186*** (-3.42)	-0.033 (-1.37)
South America	I	-0.031 (-1.88)	-0.076*** (-2.68)	-0.033 (-1.93)
	II	-0.025 (-1.43)	-0.073** (-2.36)	-0.032 (-1.63)
America	I	-0.015 (-1.63)	-0.059*** (-3.16)	-0.016 (-1.7)
	II	-0.02 (-1.51)	-0.091*** (-3.28)	-0.025 (-1.71)
East Europe	I	-0.008 (-1.44)	-0.083*** (-4.58)	-0.023** (-2.45)
	II	-0.026*** (-3.48)	-0.132*** (-4.85)	-0.043*** (-3.43)
West Europe	I	-0.002 (-0.93)	-0.028*** (-3.33)	-0.022*** (-3.63)
	II	-0.027*** (-2.98)	-0.034* (-1.74)	-0.029** (-2.57)
Europe	I	-0.011*** (-3.05)	-0.042*** (-4.89)	-0.012*** (-3.13)
	II	-0.027*** (-4.4)	-0.069*** (-3.99)	-0.03*** (-4.16)
World	I	-0.012*** (-3.71)	-0.038*** (-5.33)	-0.015*** (-3.94)
	II	-0.028*** (-4.26)	-0.166*** (-11.12)	-0.056*** (-6.38)

Note: Figures in parentheses are t-ratios. Estimated coefficients significant at 1%, 5% and 10% level are marked with '***', ** and '*', respectively.

The computed F values for tests of parametric restriction on the ECM relating to GCT are given in Table 6.5a. The parameters η_y and η_x in Table 6.5b are interpreted as the speed of adjustment coefficients which measure the speed at which the values of y_t and x_t come back to long run equilibrium levels, once they violate the long run equilibrium relationship. These parameters are of particular interest as they have important implications for the dynamics of the system. The negative sign of the estimated speed of adjustment coefficients are in accord with the convergence toward long run equilibrium. The larger the value of η, stronger is the response of the variable to the previous period's deviation from long run equilibrium, if any. As the results show, η is large for Africa (26.3%) and Central America (18.6%) compared to what it is for Western Europe (2.8%). This implies that in the case of Western Europe any deviation from long run equilibrium of the value of y_t and x_t requires a much longer time for the equilibrium to get restored. Since all the η's are statistically significant for all country-groups in both the models, a change in one variable is expected to affect the other variable through a feedback system. This implies more or less a bi-directional causal relationship between income and emission for all the country-groups. It may be pointed out here that if the *EC* term is dropped from the ECM equations (C.5) and (C.6), these equations reduce to VAR equations in *difference* which are used to test Granger causality. In Chapter 4, GCT was performed using such equations. It may be noted that since the ECM contains the *EC* term additionally, the nature of causality inferred from the estimated ECM (when the estimated coefficient of the *EC* term turns out to be statistically significant) may be qualitatively different from that obtained by using the standard GCT. This is so, essentially because whereas the GCT considers only the nature of dependence between the variables in the short run movements, the ECM-based causality takes into account the deviation from long run equilibrium as well.

In this context, it may be mentioned that since the data sets used here are essentially cross-country panel data, the problem of spatial dependence may be a possibility. As is known, the issue of spatial dependence is also important to the panel unit/cointegration tests (Breitung and Pesaran (2005), Gengenbach et al., (2004, 2005)). However in the present case, the problem of spatial dependence may not be a serious one essentially because in the present data set (per capita) emission data points are not contiguous and spread over the whole world. However in this context, an alternative set of country-groups has been formed using a satisfactory econometric rule (will be discussed later) to verify the test of panel unit root/ cointegration.

Table 6.6. Results of the exercise based on the alternative country-grouping

Country-group	I: Results of Panel Unit Root Test (IPS $\bar{t}$)			
	Without time trend		With time trend	
	income	emission	income	emission
Bottom30	-0.27	-1.66	2.32	-1.11
Middle28	0.04	-2.49**	1.03	0.41
Top30	0.47	-1.44	0.67	1.74
	II: Results of Panel Cointegration Test (IPS $\bar{t}$)			
Bottom30	-2.82**	-3.34**	-2.62**	-5.23**
Middle28	-5.02**	-4.89**	-3.40**	-4.11**
Top30	-4.38**	-4.05**	-0.29	-0.21

Note: ** denotes significant at 1% level.

As mentioned earlier, there is a possibility that the country-group-specific results reported above are confounded by the heterogeneity of the countries falling into individual country-groups formed on the basis of geographic contiguity. To examine whether it is indeed so, we have considered an alternative grouping of countries such that countries having similar short run interrelationship between income and CO_2 emission are clubbed together and repeated all the exercises discussed above for these country-groups. In what follows next, these results are reported.

Let us first briefly explain the way the alternative country-groups have been formed. We have estimated the *ECM* equation (5.6) by country, arranged the 88 countries in ascending order of value of the country-specific estimated value of β_c and then formed three country-groups comprising the countries having the lowest 30, middle 28 and highest 30 values of this parameter. The compositions of the resulting country-groups show a fair amount of geographic clustering. For example, the bottom group contains 16 African, 6 South American, 3 Asian countries along with Czechoslovakia, Yugoslavia and Turkey from Europe and Fiji and New Zealand from Oceania. The middle group contains 10 Central and South American, 9 African, 5 European countries along with India and China from Asia and Australia and Papua and New Guinea from Oceania. Finally, the top group contains 14 European, 9 Asian, 5 American (including the U.S.A.) countries and Egypt and Tunisia from Africa. The original panel data set has been recast according to this country-grouping and the econometric exercises have been redone for this recast panel data set.

The results presented in Table 6.6 are qualitatively similar to those presented earlier. For example, the panel unit root test results show that the null hypothesis of unit root cannot be rejected except for the middle country-group when a time trend is not included. The null hypothesis of no cointegration is rejected in all the cases except for the top country-group when a time trend is included. In this context, it may be recalled that, as we have noted earlier, except for North America, South America, Asia, Asia excluding Japan and Oceania, for all other country-groups the null hypothesis of no cointegration was not rejected.

6.5. Conclusion

In this chapter we have reported on a study of cointegration analysis between per capita income and CO_2 emission based on a cross-country panel data set covering 88 countries and the time period between 1960-90. The results indicate presence of cointegrating relationship between the variables concerned for all the country-groups considered except those of North America, South America, Asia, Asia excluding Japan and Oceania. The subsequent analysis based on the estimation of the error correction model done for the country-groups for which the null hypothesis of absence of cointegration could not be rejected, suggests that whereas for the country-group of Africa there is a bi-directional causality between income and emission and for the country group of Central America there is an income to emission causality, for the country-groups of Europe there seems to be a reverse causality from emission to income.

Given that temporal movements of rate of growth of both income and emission are by and large stationary and more importantly that whereas for Central America the income growth rate is autonomous or exogenous, for the European country-groups it is the emission growth rate which turns out to be exogenous, the implications of these are interesting and important from the point of view of management of the global level of CO_2 emission.

In this context, it may be mentioned that a possible means of transferring the load of emission from the developed to the developing countries is through outsourcing of emission-intensive production. In fact, as Hettige et al (1992) and Agras and Chapman (1999) have shown, the degree of openness of a country (as measured by the value of total trade relative to the country's GDP) may influence the direction of income-emission causality. International trade may affect the nature of this causality through other routes as well. For example, a highly open economy, because of its easy access to fuel through international trade, may not face the fuel supply constraint and hence the income to emission causality may not get reversed. To see the efficacy of the openness of an economy for its income-emission causality pattern, we also have done a supplementary exercise in which openness of a country-group (measured as the ratio of value of trade to GDP) is included as an exogenous explanatory variable in the income-emission relationship estimated[94]. Our results suggest that openness[95] reduces CO_2 emission in Western Europe and Europe as a whole, where as it increases emission in Africa, Central America. These results thus apparently reinforce the Hettige et al and Agras and Chapman findings mentioned above and lend support to the so-called Pollution Haven hypothesis (PHH). However, needless to mention, the country specific characteristics determine the comparative advantage, which influence the trade patterns and thereby pollution level of a country. Recently trade theorists have suggested that the interaction of relative factor abundance and environment policy may have a significant effect on the pollution level of a country. Antweiler et al., (2001) have used the interaction of openness with relative income per capita and relative capital-labour ratio as explanatory variables in their analysis. Their estimated effect of these interactions turn out to be small, suggesting perhaps that the PHH and factor endowment hypothesis (FEH) counteract and tend to offset each other (Copeland and Taylor 2004). In this context it may be mentioned that in a study of the effect of trade between the USA and China on the level of pollutions of these countries, both the PHH and FEH are rejected, which perhaps means that whether trade would explain pollution remains an unresolved puzzle (Temurshoev (2006)). In details, chapter 8 investigates on this issue by incorporating the above mentioned interaction terms into our analysis.

[94] This measure of openness is given in the Penn World Table for individual countries for each year.

[95] Hettige et al., (1992) find that toxic intensity decreases with openness of the economy, but the growth rate of the toxic intensity of manufacturing increased in the poorest countries.

Chapter 7

Carbon Dioxide Emission and Income: A Temporal Analysis of Cross-Country Distributional Patterns*

7.1. Introduction

In Chapter 6, existence of a cointegrating relationship between income and emission of CO_2 for different country-groups of the World was examined using a cross-country panel data set and it was found that such a relationship generally exists. The existence of an *equilibrium* income-emission relationship for individual country-groups suggests existence of a corresponding relationship between the inter-country disparities in the levels of income and CO_2 emission. In this chapter, we explore the relationship between the patterns of cross-country distribution of income and CO_2 emission and temporal shifts in such a relationship. To be specific, here we study how the pattern of cross-country income distribution[96] may have affected the global environment through its effect on the corresponding cross-country distribution of CO_2 emission. As we shall spell out, our primary objective here is to see whether and to what extent a change in the cross-country income distribution pattern may result in a change in the shape of the income-emission relationship and also the corresponding cross-country distribution pattern of CO_2 emission.

In the *EKC* literature environmental quality is expressed as a function of the level of income alone, ignoring the potential role that the pattern of income distribution may play in the determination of the level of environmental quality. Some recent studies, however, have brought explicitly distributional issues into the study of income-environmental quality relationship. For example, Torras and Boyce (1998) concluded that the distributional inequality of income and/or power may be an important determinant of the level of environmental degradation and a lower distributional inequality of income would result in a lower level of environmental degradation[97]. Following a public good choice approach, they

---

* This chapter is based on Coondoo and Dinda (2008).

[96] Generally, by income distribution we mean that national income of a country is distributed among the individuals within the citizens of the country. There is a great difference between the distribution within country and between countries.

[97] Scruggs (1998), however, has challenged this conclusion on both theoretical and empirical grounds.

argued that the society's decision about the level of environmental degradation would emerge on the basis of relative strengths of different interest groups as reflected by the pattern of distribution of income and social power across these groups and the extent of inequality there in. Redistribution of income may affect a *society's demand for environmental quality* through other routes as well (Magnani, (2000)). For example, a change in income distribution may bring in a new pattern of consumer demand fulfillment of which may have important environmental quality consequences (Grossman and Krueger (1995)). A more equitable income distribution may, by contributing to social harmony, also help create public opinion in favour of investment required for improving environmental quality. In this context, it may be mentioned that measures such as wider literacy, greater political liberty and civil rights that promote a more equitable distribution of income and hence of power may help improve environmental quality in poor countries. In fact, in Torras and Boyce (1998), literacy, political liberty and civil rights turned out to be better proxies for power inequality and the effect of inequality on environmental quality worked out to be stronger in poorer countries.

On the whole, studies on EKC that explicitly recognise income distribution as an explanatory factor seem to agree on at least two points, viz., (1) distributional issues are important determinants of environmental quality and (2) a lower inequality is good for the quality of environment in the long run. Thus, Kuznets' original hypothesis of an inverted-U shaped relationship between income inequality and mean level of income can be an additional channel through which economic growth may influence the environmental quality. The specific mechanism through which income inequality may affect the income-environment relationship can be the differential marginal propensities to pollute *(MPP)* of the rich and the poor.

The phenomenon of income distribution may also be relevant in a discussion of inter-country variation in the levels of pollution and the pattern of its temporal variation, as has been done in this chapter based on the cross-country time series data on per capita CO_2 emission. For example, if *MPP* is higher for poor countries than what it is for rich countries, a greater inter-country income inequality should raise the aggregate pollution level for any given level of all-country mean income. In fact, looked from this angle, an improvement of the world income distribution may result in a deterioration of environmental quality (Ravallion et al., (2000))[98]. It should, however, be mentioned here that the pattern of temporal variation of inter-country distribution of CO_2 emission is likely to be affected significantly through an altogether different mechanism – viz., divergence or convergence of the country-specific income growth rates. From that macroeconomic perspective, explanation of observed temporal variation of inter-country distribution of CO_2 emission should be attempted by bringing in factors that are likely to explain inter-country disparity in income growth rates (and hence inter-country distributional inequality of income) like, degree of openness of economies and extent of trade liberalization, diffusion of technological knowledge etc.(Barro and Sala-I-Martin 1995, chapter 12, p414 - 461). However, no such analysis has been done

[98] Some economists maintain the optimistic view that individual preferences of the rich people eventually lead to a '*virtuous circle*' relationship between rising income and environmental degradation. Empirical evidence from cross-country comparisons suggests that economic growth in poor countries entails worsening environmental outcomes. For a few environmental indicators, the evidence also suggests that the direction of the relationship is eventually reversed so that with enough growth environmental outcomes may ultimately begin to improve. The existence of such non-linearity in the cross-country relationship between environmental indicators and average incomes has implications for the relationship between income/emission inequality and environmental outcomes. Here we focus on those implications.

here and our objective here is rather humble, viz., to examine econometrically to what extent the temporal variations of mean per capita distribution of CO_2 emission level, inter-country distributional inequalities of distribution of per capita CO_2 emission and per capita income are significantly related.

Thus, in this chapter we examine three things - viz., whether the cross-country income distributional inequality of individual country-groups has any significant effect on the corresponding mean level of CO_2 emission of the country-groups, how the cross-country distributional inequalities of income and CO_2 emission are related and also how these observed distributional relationships may have changed over time. This analysis has been done mostly by applying the technique of concentration curve[99] analysis (Aitchison and Brown, 1957) to the cross-country panel data set on per capita income and CO_2 emission described in section 5.3 of chapter 5. In what follows, we explain first the methodological framework that has been used to link up the distributional inequality of income with that of the environmental variable concerned and then present the empirical results of our analysis.

The chapter is organized as follows: Section 7.2 presents the conceptual and methodological framework of the analysis. The empirical results are presented in Section 7.3. Finally, in Section 7.4 some concluding observations have been drawn.

7.2. *EKC* AND THE DISTRIBUTIONAL ISSUE

As such the statement of the EKC hypothesis makes no explicit reference to the income distribution. However, in the discussion of income-environmental quality relationship, income distribution may enter through either or both of the following two routes. First, as already mentioned, treating environmental quality as a non-excludable public good and following the public choice approach, one may argue that the observed level of environmental quality is determined by the relative powers of various interest groups of the society, where the power distribution may be closely related to income and other relevant socio-economic inequalities. Alternatively, one may regard environmental quality as a *private* consumption good and visualize the environmental quality-income relationship as an engel curve for the former. As preference for environmental quality is expected to change systematically with income, one would expect the engel curve for environmental quality to be a nonlinear one, having a negative slope at lower levels of income and a positive slope beyond a threshold income level - thus suggesting that as an item of consumption, environmental quality changes its status from a luxury/necessary to an inferior good with the rise of income. Thus, as Beckerman (1992) puts it, if someone wants "a better environment," (s)he has to "become rich." This may be true for an individual, a household, a country or a nation or the human society as a whole. In what follows, we take this latter route to examine the distributional issues involved in the income-environmental quality relationship.

Consider a society consisting of n individuals and let z denote the level of environmental quality (measured as environmental degradation on an appropriate continuous scale) desired or demanded by an individual who has an income level y . *Ceteris paribus*, let the demand function for environmental quality be

[99] See also Kakwani (1980) for a comprehensive discussion on these measures.

$$z = f(y), \tag{7.1}$$

where $f'(y)$ measures the marginal income response of environmental quality demanded. It is reasonable to expect $f'(y)>0$ for individuals having income below a threshold level y^* and $f'(y)<0$ for individuals having income above y^*. Without loss of generality, let us specify this demand function to be a polynomial in the variable *y* and write

$$z = \beta_0 + \beta_1 y + \beta_2 y^2 + \ldots\ldots \tag{7.2}$$

Next, suppose the income distribution is given by the probability density function $g(y;\theta)$, θ being a (set of) parameter(s). The aggregate or mean level of environmental quality demanded is then

$$E(z) = \int (\beta_0 + \beta_1 y + \beta_2 y^2 + \ldots.)g(y;\theta)dy = \beta_0 + \beta_1 E(y) + \beta_2 E(y^2) + \ldots\ldots \tag{7.3}$$

Thus, the mean level of environmental quality demanded depends not only on the mean income level, but also on the higher order moments of the income distribution, in general. If the demand function for environmental quality is such that it can be adequately approximated by a quadratic function of y, then the mean income and relative inequality of the income distribution together will explain the variation in the mean level of environmental quality demanded.

The inequality in the levels of environmental quality demanded across individuals that is due to the inequality of the income distribution can be analysed using the technique of *specific concentration curve analysis* as briefly explained below (Kakwani (1980)). Consider

$$G_z^*(y) = \frac{\int_0^y E(Z/v)g(v;\theta)dv}{E(Z)} = \frac{\int_0^y f(v)g(v;\theta)dv}{E(Z)} \tag{7.4}$$

which measures the cumulative share in the aggregate of environmental quality demanded of those having income up to *y*. Also consider

$$G(y) = \int_0^y g(v;\theta)dv \tag{7.5}$$

which measures the cumulative proportion of individuals having income up to *y*, $y \in (0,\infty)$. The Specific Concentration Curve *(SCC)* for the variable z is then defined implicitly as

$$\phi_z(G(y), G_z^*(y)) = 0, \tag{7.6}$$

a function that relates the cumulative share in aggregate demand for environmental quality of those having income up to *y* to the cumulative proportion of individuals having income up to *y*. The *SCC* defined above is similar in nature to the Lorenz Curve *(LC)* for *y,* which relates the share of aggregate income received by those having income up to *y* to the proportion of population having income up to *y*. Formally, the *LC* for *y* is defined implicitly as

$$\phi_y(G(y),G_y^*(y))=0, \tag{7.7}$$

where

$$G_y^*(y)=\frac{\int_0^y vg(v;\theta)dv}{E(Y)}, \tag{7.8}$$

i.e., the share in aggregate income of individuals having income up to *y*. Plotting G_y^* against G , one gets the *LC* of *y*. Similarly, plotting G_z^* against G , one gets the *SCC* of z. It may be noted that the *SCC*, as defined above, is supposed to reflect the portion of relative distributional inequality of *z* that is due to the distributional inequality of *y*.

As regards the shapes of these concentration curves, it is well known that the *LC* is a non-decreasing convex function as the slope $\frac{dG_y^*}{dG}$ is positive and non-decreasing in *y*. In the case of equal distribution, i.e., when $G_y^*(y)=G(y)$ for every *y*, the *LC* will be the 45^0 line, which is known as the egalitarian line or the line of equal distribution. The SCC is also a non-decreasing function. However, the shape of the curve depends on the nature of relationship between *z* and *y*. Let us try to explain this. For the sake of convenience, let us assume the demand function for environmental quality to be of the constant elasticity form, viz., $z=Ay^{\eta}$, where η denotes the *constant* income elasticity of demand for environmental quality. If $\eta>1$ (i.e., environmental quality is a luxury good), both *SCC* and *LC* will be non-decreasing convex functions, but as $G_z^*<G_y^*$ for every level of *y*, the *SCC* will lie below the *LC*. If $0<\eta<1$ (i.e., environmental quality is a necessary good), then also both *SCC* and *LC* will be non-decreasing convex functions, but the *SCC* will lie above the *LC* as $G_z^*>G_y^*$ for every level of y in this case. Thus, in this case the *SCC* will lie between the egalitarian line and the *LC*. Finally, if $\eta<0$ (i.e., environmental quality is an inferior good), then the *SCC* will be a non-decreasing concave function and lie above the egalitarian line for all levels of *y*.

Under the *EKC* hypothesis, initially *z* should fall as *y* increased and beyond the turning point income level, *z* should rise with *y*. In terms of engel elasticity this means that *z* is a luxury or necessary below the turning point income level (y^*, say) and an inferior good above y^*. Thus, if *EKC* hypothesis holds, the *SCC* for *z* will intersect the egalitarian line from below at the turning point income level.

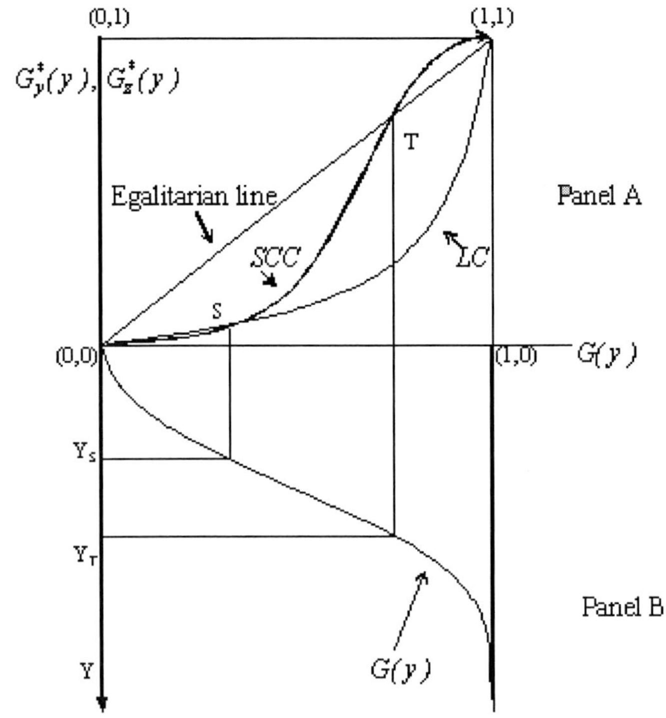

Figure 7.1. Lorenz Curve of income and Specific Concentration Curve for emission.

Figure 7.1 gives a diagrammatic presentation of the *LC*, *SCC* and their interrelationship. The upper part of the figure gives the *LC* and the *SCC*. The *SCC* intersects the *LC* from below at the point S. Thus, at income levels below Y_S , z is a luxury good. Next, at point T the SCC intersects the egalitarian line from below and goes above that line. So, in the income interval (Y_S , Y_T), z is a necessary good, T being the turning point and Y_T being the corresponding turning point income level.

It may be mentioned here that if the income distribution changes from year to year, both *LC* and *SCC* will shift correspondingly, there will be a temporal drift in the observed income-emission relationship and so the turning point income level will also shift. A related issue of interest is the extent to which a change of the income distribution will affect the corresponding distribution of environmental quality demanded. This issue assumes importance in a cross-country set up for the following reason. If it is supposed that the level of CO_2 emission (or for that matter any kind of environmental degradation, in general) is correlated with the income level, then at a specific time point the cross-country distribution of emission level will be determined by the corresponding cross-country income distribution. More importantly, a change in the relative inequality of the cross-country income distribution will affect both the global (i.e., the aggregate or the average level of emission or environmental degradation, as the case may be) level of emission and the cross-country distributional inequality of emission. To examine this, one may use the summary measures of relative inequality based on the *LC* and the *SCC*, viz., the Lorenz Ratio *(LR)* and Specific Concentration Ratio *(SCR)*. These measures are defined below:

$$LR = 1 - 2\int_0^1 G_y^*(y)dG(y) \text{ and } SCR = 1 - 2\int_0^1 G_z^*(y)dG(y). \quad (7.9)$$

It may be mentioned that whereas $LR \in (0,1)$ with $LR = 0$ and $LR = 1$ signifying complete equality and complete inequality of income distribution, respectively, $SCR \in (-1,1)$. SCR assumes the value -1 for an inferior good, all of which is consumed by the poorest person and it assumes the value 1 for a luxury good, all of which is consumed by the richest person of the society. In case of both *LR* and *SCR*, however, a rise in the value of the measure signifies an increase in the distributional inequality.

Table 7.1. Country-group-specific Summary Statistics

Country-group	Variables	Mean	Median	Maximum	Minimum	Standard Deviation
World						
	Income	3482.94	3530.98	4457.44	2402.60	619.19
	Emission	0.96	0.99	1.07	0.77	0.09
	LR	0.57	0.57	0.59	0.55	0.01
	SCR	0.53	0.54	0.57	0.51	0.02
Africa						
	Income	1367.78	1491.59	1702.69	947.65	259.87
	Emission	0.35	0.38	0.46	0.21	0.09
	LR	0.29	0.28	0.33	0.25	0.02
	SCR	0.25	0.25	0.30	0.21	0.03
America						
	Income	8163.65	8411.81	9773.19	6022.21	1161.04
	Emission	2.52	2.49	2.86	2.21	0.18
	LR	0.34	0.34	0.37	0.32	0.01
	SCR	0.34	0.34	0.37	0.32	0.01
Asia						
	Income	1457.97	1424.36	2251.55	853.14	426.59
	Emission	0.35	0.37	0.54	0.18	0.11
	LR	0.39	0.39	0.43	0.27	0.03
	SCR	0.32	0.34	0.37	0.15	0.06
Europe						
	Income	7976.71	8217.68	10920.30	4891.66	1747.64
	Emission	2.02	2.04	2.32	1.56	0.20
	LR	0.19	0.19	0.23	0.18	0.01
	SCR	0.13	0.13	0.18	0.11	0.01

7.3. The Results

Using this cross-country panel data set described in section 5.3 of chapter 5 and applying the *LC* and *SCC* techniques explained above, we have obtained the *LC* and *SCC*, corresponding *LR* and *SCR* and also mean income and mean emission level for each of the years 1960-1990 for five country-groups consisting of countries of Africa, America (North, Central and South America pooled together), Asia, Europe and finally the World, taken as a whole[100].

Let us first examine the country-group-specific summary statistics relating to income, CO_2 emission, *LR* of income and *SCR* of CO_2 emission presented in Table 7.1. It may be observed that average emission and income levels of America and Europe are quite high and much above the global average, whereas the corresponding figures for Asia and Africa are lower and below the global average level. This confirms the commonly held view that developing countries contribute much less to the global CO_2 emission compared to their developed counterparts. Average inequality of income[101] *(LR)* is lowest for Europe (19.39 per cent) and highest for Asia (38.56 per cent). Understandably, the mean income inequality for the World as a whole is much higher (56.61 per cent) than these figures. The average *SCR* is also low for Europe (13.06 per cent) and high for America (33.82 per cent) and it is the highest for the World (53.32 per cent). On the while, *LR* and *SCR* are low for the developed world and high for the developing world. It may also be noted that the country-group of Africa has both the lowest mean level and the lowest inequality of income and emission. The standard deviations of these inequality measures for the period under study are also given in the last column of Table 7.1 as a summary measure of variation.

7.3.1. EKC and LR

As already mentioned, environmental quality may get affected considerably by relevant socio-economic inequalities. Consider, for example, equation (6.3). If *E(z)* is taken as a measure of mean level of emission for a country-group, then the mean level of income and the *LR* of the country-group should be the basic explanatory variables of the emission-income relationship if the country-level engel curves for environmental quality demanded are quadratic functions of country-level mean income.

In order to examine whether *LR* indeed has any significant effect on the mean level of emission, we have done the cointegration exercise using Johansen Procedure separately for each country-group, based on the country-group-specific time series data set of yearly mean per capita emission, mean income and *LR*. The results of this exercise are summarised in Table 7.2.

It may be noted first that mean income, mean emission and *LR* are cointegrated for all the country-groups. Except for the country-group of Europe, for all the other country-groups there is one statistically significant cointegrating relationship and for the country-group of Europe two significant cointegrating vectors are estimated. For this country-group, the effect of mean income on mean emission is non-significant, whereas for all the other country-groups the effect of a rise in mean income, given the *LR*, on the mean level of emission is positive. The effect of *LR* on the mean emission level, on the other hand, is significant for all the country-groups except Africa. However, whereas for the country-groups of America and

[100] The country-group-specific concentration analyses have been done by taking into account the population size of the constituent countries.

[101] It may be noted that here LR and SCR being based on the data on country-specific mean values of the variables concerned, give the "between-country group" relative inequality measures.

Europe a rise in *LR* is expected to decrease the mean emission level, for the country-group of Asia and the World as a whole, the effect seems to be in the opposite direction.

Table 7.2. Estimated cointegrating vectors relating mean emission, mean income and LR by country-group

Country-group	Number of cointigrating vectors estimated	Elements of the normalized estimated cointegrating vector			
		Mean emission	Mean Income	LR	Intercept
World	1	1	$-2.4x10^{-4}$ ** (-6.05)	-9.09** (-3.56)	5.26** (3.2)
Africa	1	1	$-2.7x10^{-4}$ ** (-7.71)	0.28 (0.51)	-0.11 (-0.52)
America	1	1	$-1.5x10^{-4}$ * (-2.43)	17.46* (2.39)	-7.58** (-3.2)
Asia	1	1	$-2.6x10^{-4}$ ** (-28.06)	-0.78** (-6.08)	0.30** (6.26)
Europe	2	1	$-5.2x10^{-5}$ (-1.6)	14.35** (4.6)	-4.66** (-6.2)
		1	0	15.86** (9.35)	-5.19** (-14.72)

Note: Figures in parentheses are the t-ratios. One and Two asterisks (*, **) indicate that an estimated coefficient is significant at 5% and 1% level, respectively.

A plausible explanation of this result may be as follows. The richer countries of America and Europe have significantly smaller *marginal propensity to emit* (*mpe)*, so that when *LR* rises and the income distribution shifts in favour of these countries, the mean emission level tends to decline. Following the same argument, the richer countries of Asia may not be having significantly smaller *mpe* compared to their poorer counterparts and therefore a shift of the income distribution in their favour tends to raise the mean emission level. These results are in line with those of Holtz-Eakin and Selden (1995), World Bank (1992), Shafik (1994) and Sengupta (1997).

Finally, as the results for the country-group of Asia tend to suggest, the net effect of a rise in mean income with a concurrent decrease in the *LR* may very well be a decrease in the mean level of emission depending on the magnitudes of the changes in the mean income and the *LR*. This is in line with the view that an equalizing income growth across countries may help contain the level of emission globally. This may have some policy significance so far as the management of aggregate level of emission in Asia is concerned.

7.3.2. SCR and LR

As already stated, in the present exercise mean emission level and mean income level have been found to be cointegrated for all the country-groups considered. Since the cointegrating relationship binds together the temporal movements of mean income and emission levels, it is imperative that the cross-country distributional patterns of emission and income would also be related to each other. A graphical examination of the temporal movements of *SCR* and *LR* over the period 1960-90 for different country-groups also confirmed this conjecture. For a closer examination we therefore have performed a cointegration analysis involving the three variables *SCR, LR* and mean income level separately for each of the country-groups.

The results of this cointegration analysis are presented in Table 7.3. It may be mentioned first that for all the country-groups at least one statistically significant cointegrating vector is found. Whereas the number of cointegrating vectors for the country-groups of America, Europe (and also the World as a whole) is found to be one, the number of such vectors for Africa and Asia are found to be two and three, respectively. For both Africa and Asia one of the estimated cointegrating vectors does not involve the *SCR* (implying thereby that this vector defines a relationship between mean income level and *LR* only). Of the remaining two cointegrating vectors for Asia, one involves all the three variables while the other involves *SCR* and *LR* but not mean income.

Table 7.3. Estimated cointegrating vectors relating SCR, mean income and LR by country-group

Country-group	Number of cointegrating vectors estimated	Elements of the normalized estimated cointegrating vector			
		SCR	Mean Income	LR	Intercept
World	1	1	$-0.97x\ 10^{-5}$ **	-1.84	0.54**
			(-3.23)	(-1.41)	(6.58)
Africa	2	1	$-5.0x10^{-5}$ **	-1.05**	0.12**
			(-17.86)	(-31.76)	(10.63)
		0	1	18724.02*	-7307.69*
				(1.92)	(-2.43)
America	1	1	$0.12x10^{-5}$ **	-1.02**	$7.7x10^{-4}$
			(8.88)	(-84.2)	(0.19)
Asia	3	1	$-7.5x10^{-5}$ **	-2.15**	0.65**
			(-6.88)	(-15.19)	(10.59)
		1	0	-1.08*	0.088
				(-2.41)	(0.48)
		0	1	14228.68*	-7446.02**
				(2.34)	(-2.96)
Europe	1	1	$0.25x10^{-5}$ *	-0.32	-0.092
			(1.78)	(-1.34)	(-1.66)

Note: Figures in parentheses are the t-ratios. One and Two asterisks (*, **) indicate that an estimated coefficient is significant at 5% and 1% level, respectively.

Let us next consider the nature of dependence of *SCR* on the remaining variables as reflected by the relevant estimated cointegrating vectors. The patterns of dependence of *SCR* on *LR* and mean income for the country-groups of America and Asia are qualitative similar to those of Europe and Africa, respectively. Thus, for example, for America and Europe, a rise in mean income level, given *LR*, would reduce *SCR* and a rise in *LR*, given the mean income, would raise *SCR*. The partial effect of *LR* on *SCR* is however numerically stronger for America compared to that for Europe (which may be primarily due to much greater heterogeneity of the countries falling into the country-group of America). In case of Africa and Asia, on the other hand, a rise in mean income level, given *LR*, would bring down *SCR* and a rise in *LR*, given the mean income, would raise *SCR*. Briefly thus, whereas for every country group *LR* affects the corresponding *SCR*, the nature of the effect is qualitatively different for the developing and developed country-groups.

7.3.3. Evidence on Existence of a Turning Point

As we have already explained, if the *SCC* intersects the egalitarian line from below, such an intersection signifies that emission or environmental degradation becomes an inferior good at income levels above the one that corresponds to this point of intersection. This latter income level is known as the *turning point* in the *EKC* literature. In terms of the *EKC*, the emission level start declining with income along the *EKC* once the *turning point* income level has been crossed. In other words, existence of a turning point on an empirical *EKC* is indicative of environmental improvement[102]. We have investigated the existence of turning point for individual country-groups by checking numerically if the empirical *SCC* intersected the egalitarian line. This exercise has been done for all the country-groups for each of the years 1960-90.

Let us summarise the results of the exercise exploring the existence of turning point. Of all the country-groups, evidences of turning point are found only for that of Europe. The SCC for the country-group of Europe is seen to cross the egalitarian line from the year 1966 onward[103]. Interestingly, the turning point income level is seen to increase monotonically over time. The European countries for which the mean per capita income level is generally found to be higher than the turning point income level in most years are observed to be Denmark, Luxembourg, the Netherlands, Sweden, Switzerland and West Germany. These results, thus, seem to provide empirical evidence in support of the *EKC* hypothesis for CO_2 emission in case of the countries of Europe alone. Further, the fact that the estimated turning point income level (measured at constant prices) is observed to be rising monotonically over time is perhaps suggestive of the fact that even the rich countries of Europe find it hard to bring down their CO_2 emission levels. Finally, it should be mentioned that these observed turning point income levels are observed within the sample income levels, which contradicts the findings of Holtz-Eakin and Selden (1995)[104] for CO_2 emission.

[102] In the *EKC* literature, evidences of two, rather than one, turning points of empirical *EKC* have been reported (see, e.g., Sengupta (1997), Grossman and Krueger (1995)). The second turning point income level (which is greater than the first one) signifies the beginning of a new phase of rising environmental degradation required for improving further the already-reached high income level.

[103] The income level corresponding to the point of intersection of the *SCC* and the egalitarian line is calculated by interpolation.

[104] Holtz-Eakin and Selden (1995) estimated the turning point income level for the World as a whole to be $34000 approximately, a value that fell beyond the sample range of income.

7.4. CONCLUSION

In this chapter we have examined how the mean level of per capita CO_2 emission and its distributional inequality is related to the corresponding mean income level and the distributional inequality of income at the level of country-group based on a cross-country panel data set. Treating environmental quality demanded as a *private good* (and not as a public good as done in most other studies), we have used the technique of Lorenz and specific concentration curve analysis as the basic analytical framework to argue that a measure of relative distributional inequality of income should used as an explanatory variable in the *EKC* relationship along with the mean income level. In the empirical exercise, we have used the Johansen's cointegration analysis technique to explore existence of statistically significant cointegrating vector(s) relating mean level of emission and *SCR* of emission to mean income level and *LR* of income.

The empirical results confirm that the pattern of cross-country income distribution has significant effect on the mean level of emission for all the country-groups considered. More importantly, it is found that whereas for the richer country-groups of Europe and America an equalizing redistribution of income would bring down the mean emission level, the opposite is the case for the poorer country-groups of Asia and Africa. A significant effect of *LR* on the *SCR* for emission is also observed. In this case also, the results for the developed country-groups are found to be qualitatively different from those of their poorer counterparts. Finally, evidences in favour of existence of turning point income level on the empirical *EKC* based on the *SCC* have been found for the country-group of Europe alone for the period 1966 onwards.

One should find the basic analytical framework of concentration curve analysis that we have used here quite novel and convenient for the purpose of *EKC* analysis. One would also find the empirical results and the conclusions based on them that we have drawn sensible and interesting. However, the chapter is not free of limitations. For example, since the analysis has been designed at the level of a country-group, both the mean emission level and the cross-country disparity of emission level is likely to be significantly affected by the size of international trade that individual countries are engaged in. In fact, the so called *Pollution Haven Hypothesis* asserts that many rich developed countries are more and more outsourcing their requirements of emission-intensive material goods from poorer developing countries and thereby shifting bulk of their emission to the latter countries. When this happens, the cross-country disparity reflected in the *SCC* and *SCR* may not be as closely linked up to the cross-country disparity of income reflected by the LC or LR of income. Needless to mention, one should bring in an appropriate measure of *openness* of the individual countries in to the analysis to identify the pure partial effect of income distribution on the emission level.

Chapter 8

IMPACT OF GLOBALIZATION ON ENVIRONMENT: HOW DO WE ANALYZE IT?

8.1. INTRODUCTION

Economic development through rapid industrialization and growing environmental consciousness together has generated a heated debate on how economic development/growth is linked with environment. The linkage of environmental quality with economic development evoked much discussion in the last decade (i.e., 1990s). The World Development Report (World Bank 1992) presented cross-sectional evidences on the relationship between different indicators of environmental quality and per capita national income across countries. Other studies (e.g., Grossman and Krueger, 1991; Selden and Song, 1994; Rothman, 1998; Suri and Chapman, 1998; etc) documented an inverted U-shaped relationship between environmental degradation and income. The common point of all these studies is the assertion that environmental degradation increases initially, reaches a maximum level and after that declines as an economy develops. This systematic inverted-U relationship has been termed as the *Environmental Kuznets Curve* (*EKC*) following the work of Kuznets (1955), who postulated a similar relationship between income inequality and economic development. The *EKC* relates to the issue of the impacts of economic growth or development on the environment of *a country*. Now automatically one important question arises whether cross border integration (i.e., globalization) helps or hurts this process. This is indeed the primary motivation for this study.

Globalization is reflected in rising trade shares in world GDP. So, directly the activity of trading itself generates pollution. However, globalization, by increasing competition for investment, may trigger the environmental *race to bottom* (Wheeler (2000)). Poor economies may be able to improve their environmental quality as investment raise their income levels. Thus, globalization may facilitate pollution reduction. In fact, *the bottom* rises with economic growth. Tisdell (2001) points out that globalization can be a driving force for global economic growth. Yet opinion is divided about the benefits of this process. The global economy raises the issue of potential conflicts between two powerful current trends – viz., the worldwide acceptance of market oriented economic reform process on the one hand, and environmental protection on the other.

The aim of this paper is to explore the impact of globalization on the environment. Next question is how we measure, interpret or analyze it. For this purpose, we consider pollution at per capita level, pollution intensity and relative change of pollution for the developed (OECD) and under developed (Non-OECD) country groups and the world as a whole. All economic activities require energy, which generates pollutions. The source of energy is mostly fossil fuels that definitely release CO_2 emission, which is the main culprit of global warming and climate change. Therefore, in this study we use CO2 emission which is measured in tons. This analysis has been done mostly by applying the technique of panel data analysis to a cross-country panel data set on per capita income and CO_2 emission. The observed results suggest that the impact of globalization on environment heavily depends on the characteristics of the economy. The basic characteristics of a country and its dominating comparative advantage determine how trade liberalization influences its sectoral composition and consequently environmental outcomes.

This chapter is organized as follows: Section 8.2 presents the literature; Section 8.3 describes the data set; the empirical results are presented in Section 8.4 and finally, Section 8.5 concludes.

8.2. Literature

Globalization raises the trade volume of countries as well as trade share in the world income and consequently pollution level increases. Environmental quality could decline through the *scale effect* as increasing trade volume (especially export) would expand the size of the economy and thereby increasing the extent of pollution. Thus, trade might be a cause of environmental degradation, *ceteris paribus*. Many economists have long argued that trade is not the root cause of environmental damage (Birdsall and Wheeler 1993, Lee and Roland-Host 1997, Jones et al., 1995). However, free trade has the contradictory impacts on environment, both increasing pollution and motivating reductions in it. Antweiler et al., (2001) and Liddle (2001) argue that trade may be good for environment. Trade may improve the environmental quality through *technological effect*. As income rises through trade, environmental regulation is tightened that spurs pollution reducing innovation. And as trade relates one country with international communities, one underdeveloped economy may rely on technology transfer through foreign direct investment that may reduce pollution.

8.2.1. Pollution Haven (PH) Effect

It should be mentioned that a polluting activity in a high-income country normally faces higher regulatory costs[105] than its counterpart in a developing country (Mani and Wheeler (1998)). Under these circumstances the pollution intensive industries will have a natural tendency to migrate to countries with weaker environmental regulations (Copeland and Taylor (1995)). This is referred to the *Pollution Haven (PH) effect*. The *PH effect* predicts

[105] This creates an incentive for at least some highly polluting industries to relocate. The firms are relocated to low-income countries with weak environmental regulation. Rising capital outflows force governments in high-income countries to begin relaxing environmental standards.

that, under free trade, multinational firms will relocate the production of their polluting goods to developing countries, taking advantage of the low environment monitoring in these countries. Developing countries will develop a comparative advantage in pollution-intensive industries and become 'haven' for the world's polluting industries[106]. Thus, developed countries are expected to benefit in terms of environmental quality from trade, while developing countries will lose (See, Figure 1). In other words, the *PH effect* basically suggests that countries having stricter environmental standard will lose all the *dirty industries* and poor countries (i.e., those having poorer environmental standard) will get them all.

8.2.2. Displacement Hypothesis

It should be mentioned that the differences in the consumer preferences for a cleaner environment in rich and poor countries are also important. It is observed that changes in the structure of production in developed economies are not accompanied by equivalent changes in the structure of consumption. This could be explained by the *displacement hypothesis*. As Rothman (1998) speculates, what appears to be an improvement in environmental quality may in reality be an indicator of increased ability of consumers in wealthy nations to distance themselves from the environmental degradation associated with their consumption[107]. Hettige et al (1992) observe that toxic intensity grew rapidly in high-income countries during 1960s and this pattern was sharply reversed during 1970s and 1980s, after the advent of stricter environmental regulations in OECD countries. Concurrently, toxic intensity in LDC manufacturing grew quickly. Lucas et al (1992), Low and Yeats (1992) also confirm this *displacement hypothesis*.

8.2.3. Pollution Haven Hypothesis (PHH)

The inverted-U curve might partly be the result of changes in international specialization under which poor countries engage in *dirty* and energy intensive production while rich countries specialize in *clean* and service intensive production, without effectively any change in the consumption patterns. In this context, globalization, which reduces transport costs and/or trade barriers, would shift investment and production of dirty goods to less developed countries. As a result, globalization would lead to an increase in the pollution content of imports by developed countries, in conformity with what has been called the "*Pollution Haven Hypothesis*", and to under developed countries increase in the production of dirty goods (See, Bommer (1999), Cole (2003, 2004)). Agras and Chapman (1999) and Suri and Chapman (1998) analyse the composition of international trade and observe that manufacturing goods exporting countries tend to have higher energy consumption. They find the poor and rich countries to be net exporters and net importers of pollution-intensive goods, respectively; that support the PHH.

[106] Basically, the *PH effect* refers to the possibility that polluting industries concentrate in developing countries with low environmental standards.

[107] The mechanisms through which such distancing takes place may include both moving sources of pollution away from the people and moving people away from pollution sources. Gawande et al., (2000, 2001) provide evidence that migration is a contributing factor behind the *EKC* especially for US hazardous waste.

8.2.4. Factor Endowment Hypothesis (FEH)

On contrary to the *PHH*, the *factor endowment hypothesis* (*FEH*) asserts that in free trade the differences in endowments (or *technology*) determine trade between two countries. The *FEH* suggests that the capital abundant country exports the capital-intensive goods that stimulate its production and thereby raising pollution in the capital abundant country. The effects of trade on the environment depend on the comparative advantages enjoying a country. Under this view capital-abundant countries tend to export capital-intensive goods, regardless of differences in environmental policy (Copeland and Taylor 2004). According to the FEH[108] polluting industries will concentrate in affluent countries, which also tend to be capital abundant. This is because polluting industries are typically also capital intensive, and thus affluent capital-abundant countries have a comparative advantage in these industries (Copeland and Taylor 2004).

8.2.5. Joint Effect

In this context, it should be noted that the differences in environmental policy and differences in factor endowments might jointly determine the comparative advantage in trade. Antweiler et al., (2001) and Copland and Taylor (2004) provide a possible theoretical model with empirical support using pollution concentration data. Frankel and Rose (2005) analyze the endogenous problem and impact of globalization on environment for given income level. It is clear that effects of trade liberalization on environmental quality depend on, among other factors, jointly by differences in pollution policy and differences in factor endowments, which leads to two competing theories in question. This study adds some fuel on going debate providing some evidences and insights of policy.

8.3. The Data and Methodology

In the present exercise we have used annual per capita real GDP *(PCGDP)*, capital per worker (capital–labour (i.e., *K/L)* ratio), trade intensity and annual per capita CO_2 *(PCCO2)* emission as the measure of the income, *K/L* ratio, openness and pollution variable, respectively. It should be noted that globalization is the proxy measurement in terms of openness, which is measured as export plus import to GDP. This openness is also termed as trade intensity. The basic country-level time series data *PCGDP* (expressed in 1985 international prices, i.e., PPP dollars) for the period 1965-1990 were taken from the RGDPCH series of the Penn World Table (Mark 5.6) available at the web site http://www.nber.org/pwt5.6. As mentioned, for the present exercise we have used cross-country panel data on *Openness*, capital per worker (*K/L*), *PCGDP* and other economic variables compiled by Summers and Heston (viz., the Penn World Table 5.6). Corresponding

---

[108] Under free trade the capital abundant country exports the capital-intensive (dirty) goods, which stimulates its production, thus raising pollution in the capital abundant country. Conversely, pollution falls in the capital-scarce country as a result of contraction of the production of pollution-intensive goods, since there is no

panel data set on *PCCO2* (measured in metric tons) was obtained from the web site of Carbon Dioxide Analysis Information Center (CDAIC), Oak Ridge National Laboratory of the U. S. A. Combining these data sets, we compiled a panel data[109] set of annual observations on income, *K/L* ratio, openness and emission covering 54 countries for the time period 1965 - 1990. For the purpose of the exercise, we grouped the countries into 3 major country groups, viz., developed (OECD), less developed (Non-OECD) and the world as whole.

The empirical exercise has been done separately for each of these country groups based on these panel data sets for the individual country groups. Panel data analyses offer different ways to deal with the possibility of country-specific variables. Fixed Effect (FE) model is a suitable estimation approach that treats the level effects as constants, whereas Random Effect (RE) model is suitable to capture the level effect as variable. It should be mentioned that RE model treats the level effects as uncorrelated with other variables, while FE model does not. In this analysis we estimate both FE and RE models. Now the estimating equations are

$$E_{it} = \alpha + \beta_1 \ln Y_{it} + \beta_2 (\ln Y_{it})^2 + \beta_3 \ln KL_{it} + \beta_4 (\ln KL_{it})^2 + \beta_5 \ln P_{it} + \lambda OP_{it} + \theta t + u_{it} \tag{8.1}$$

$$\ln E_{it} = \alpha + \beta_1 \ln Y_{it} + \beta_2 (\ln Y_{it})^2 + \beta_3 \ln KL_{it} + \beta_4 (\ln KL_{it})^2 + \beta_5 \ln P_{it} + \lambda OP_{it} + \theta t + u'_{it} \tag{8.2}$$

$$\begin{aligned} E_{it} &= \alpha + \beta_1 \ln Y_{it} + \beta_2 (\ln Y_{it})^2 + \beta_3 \ln KL_{it} + \beta_4 (\ln KL_{it})^2 + \beta_5 \ln P_{it} + \lambda OP_{it} + \theta t \\ &+ \eta_1 (RKL_{it} * OP_{it}) + \eta_2 (RKL_{it}^2 * OP_{it}) + \eta_3 (RI_{it} * OP_{it}) + \eta_4 (RI_{it}^2 * OP_{it}) + v_{it} \end{aligned} \tag{8.3}$$

$$\begin{aligned} \ln E_{it} &= \alpha + \beta_1 \ln Y_{it} + \beta_2 (\ln Y_{it})^2 + \beta_3 \ln KL_{it} + \beta_4 (\ln KL_{it})^2 + \beta_5 \ln P_{it} + \lambda OP_{it} + \theta t \\ &+ \eta_1 (RKL_{it} * OP_{it}) + \eta_2 (RKL_{it}^2 * OP_{it}) + \eta_3 (RI_{it} * OP_{it}) + \eta_4 (RI_{it}^2 * OP_{it}) + v'_{it} \end{aligned} \tag{8.4}$$

Where E_{it} is the pollution measured in country i at time t; similarly *Y, KL, P, OP, RI, RKL* and *t* denote income per capita, capital labour ratio, population, openness, relative income, relative capital ratio and time trend, respectively.

8.4. Results

Following EKC analysis in globalized world, we first estimate equations (8.1) – (8.2). Later we introduce more complex model and estimate equations (8.3) – (8.4), which include the openness interaction with country characteristics. The conditioning the impact of openness on country characteristics is the important determining factors through which trade affects the pollution intensity of national output.

---

comparative advantage of producing polluting goods in the developing world (Temurshoev 2006, Liang 2006, Mukhopadhyay et al 2005).

109 see, Coondoo and Dinda (2002)

Table 8.1. Estimated Impact of Globalization on CO2 emission level

Variable	World		Non-OECD		OECD	
	FE	RE	FE	RE	FE	RE
	Estimated Coefficient	Estimated Coefficient	Estimated Coefficient	Estimated Coefficient	Estimated Coefficient	Estimated Coefficient
Log(income)	-3.462** (-4.47)	-2.109** (-2.95)	-1.79** (-9.04)	-1.879** (-9.65)	9.166 (1.94)	7.015 (1.57)
Log(income)sq.	0.271** (5.69)	0.182** (4.17)	0.138** (11.09)	0.144** (11.74)	-0.386 (-1.47)	-0.293 (-1.19)
Log(K per labour)	0.739 (1.94)	1.114** (3.03)	-0.32** (-3.54)	-0.303** (-3.4)	-3.349 (-1.37)	1.064 (0.48)
Log(K per labour)sq.	-0.04 (-1.82)	-0.068** (-3.26)	0.027** (4.86)	0.024** (4.55)	0.165 (1.31)	-0.071 (-0.62)
Log(population)	0.858** (4.81)	-0.099 (-1.59)	0.007 (0.10)	-0.033* (-1.96)	2.372** (3.61)	-0.434** (-3.04)
OPEN	**$-6.8x10^{-3}$** **	**$-7.24\ x10^{-3}$** **(-6.49)**	**$-9.75\ x10^{-5}$** **(-0.29)**	**$-4.54\ x10^{-4}$** **(-1.50)**	**-0.02**** **(-6.1)**	**-0.017**** **(-5.84)**
Time trend	**(-5.7)** -0.025**	0.001** (3.1)	-0.002 (-1.21)	0.0001 (0.34)	-0.04** (-3.42)	-0.002 (-1.57)
Constant	(-5.02) -	2.44 (1.09)	-	7.39** (10.79)	-	-34.75** (-2.91)

Note: Figures in parentheses are t-values. * and ** denote 5% and 1% level of significance, respectively.

Table 8.2. Estimated Impact of Globalization on CO2 emission intensity (per dollar)

Variable	World		Non-OECD		OECD	
	FE	RE	FE	RE	FE	RE
	Estimated Coefficient	Estimated Coefficient	Estimated Coefficient	Estimated Coefficient	Estimated Coefficient	Estimated Coefficient
Log(income)	4.1E-04** (4.37)	6.4E-04** (7.38)	2.4E-04** (3.87)	2.3E-04** (3.71)	2.9E-03** (5.72)	2.5E-03** (5.19)
Log(income)sq.	-2.6E-05** (-4.49)	-4.1E-05** (-7.74)	-1.7E-05** (-4.46)	-1.7E-05** (-4.29)	-1.7E-04** (-5.95)	-1.5E-04** (-5.48)
Log(K per labour)	1.3E-04** (2.71)	1.9E-04** (4.22)	2.6E-05 (0.91)	2.7E-05 (0.98)	-4.2E-04 (-1.59)	-1.3E-05 (-0.05)
Log(K per labour)sq.	-5.5E-06* (-2.04)	-1.0E-05** (-4.04)	7.8E-07 (0.45)	5.0E-07 (0.29)	2.4E-05 (1.76)	2.7E-06 (0.21)
Log(population)	1.5E-04** (6.83)	-7.5E-06 (-1.05)	2.6E-05 (1.26)	1.6E-06 (0.28)	1.8E-04* (2.51)	-5.7E-05** (-4.09)
OPEN	**-4.5E-07**** **(-3.06)**	**-5.8E-07**** **(-4.25)**	**3.3E-07**** **(3.14)**	**2.5E-07**** **(2.58)**	**-3.2E-06**** **(-9.18)**	**-2.5E-06**** **(-8.09)**
Time trend	-4.2E-06** (-6.92)	1.6E-07** (4.55)	-8.6E-07 (-1.51)	-2.8E-08 (-0.46)	-1.5E-06 (-1.15)	-1.7E-07 (-1.42)
Constant	-	-3.1E-03** (-11.57)	-	-9.0E-04** (-4.17)	-	-9.9E-03** (-7.63)

Note: Figures in parentheses are t-values. * and ** denote 5% and 1% level of significance, respectively.

Table 8.3. Estimated Impact of Globalization on Relative change of CO2 emission

Variable	World		Non-OECD		OECD	
	FE	RE	FE	RE	FE	RE
	Estimated Coefficient	Estimated Coefficient	Estimated Coefficient	Estimated Coefficient	Estimated Coefficient	Estimated Coefficient
Log(income)	2.32**	3.218**	2.548**	2.407**	10.606**	9.584**
	(7.13)	(10.66)	(6.46)	(6.17)	(8.09)	(7.72)
Log(income)sq.	-0.089**	-0.147**	-0.114**	-0.105**	-0.54**	-0.487**
	(-4.47)	(-8.02)	(-4.58)	(-4.27)	(-7.4)	(-7.12)
Log(K per labour)	0.798**	1.031**	0.301	0.339	-1.422*	0.120
	(4.99)	(6.65)	(1.67)	(1.90)	(-2.09)	(0.19)
Log(K per labour)sq.	-0.034**	-0.054**	0.0006	-0.0046	0.069*	-0.0117
	(-3.7)	(-6.1)	(0.05)	(-0.42)	(1.97)	(-0.37)
Log(population)	0.567**	0.016	0.418**	-0.006	0.972**	-0.069
	(7.57)	(0.57)	(3.17)	(-0.17)	(5.33)	(-1.76)
OPEN	**9.3E-04**	**-7.7E-05**	**3.5E-03****	**2.3E-03****	**-4.4E-03****	**-3.4E-03****
	(1.86)	**(-0.16)**	**(5.27)**	**(3.77)**	**(-4.9)**	**(-4.21)**
Time trend	-0.017**	0.0008**	-0.014**	-0.0001	-0.01**	-0.0006
	(-7.99)	(4.99)	(-3.92)	(-0.35)	(-3.1)	(-1.92)
Constant		-22.40**		-16.07**		-44.99**
		(-23.82)		(-11.68)		(-13.56)

Note: Figures in parentheses are t-values. * and ** denote 5% and 1% level of significance, respectively.

Tables 8.1- 8.3 present the estimated results of the impact of globalization on EKC using data of emission level, emission intensity and relative change of emission, respectively. Our central focus is, λ, the coefficient of openness (i.e., proxy of globalization). For per capita CO_2 emission level, Table 8.1 shows that the coefficient of openness is negative for all three groups and statistically significant except Non-OECD country group. This suggests that globalization helps to reduce pollution in developed country not in developing country. Thus, empirical findings of Table 8.1 support the PH effect, not the PHH. In case of CO_2 emission intensity per dollar (Table 8.2) and relative change of CO_2 emission per capita (Table 8.3), the coefficients of openness are significantly negative in OECD (and the World) but positive in Non-OECD country group. Thus, unambiguously globalization helps developed countries to reduce emission (or pollution) while it increases in under developed country. The results (Tables 8.2- 8.3) point out that globalization promotes to increase emission intensity and relative change of emission and thereby liberalization or openness hurts the environment of developing country while improves the environment of developed country. Thus, these empirical findings of Tables 8.2 – 8.3 support the *pollution haven hypothesis (PHH)*. These findings also support the earlier studies (Low and Yeats (1992), Agras and Chapman (1999) and Suri and Chapman (1998) etc.). These findings suggest that globalization forces developed countries to specialize relatively clean (or less pollution-intensive) goods production while under developed countries to specialize to produce dirty (or more and more pollution-intensive) goods (See, Figure 8.1).

Table 8.4. Estimated Interaction Results on CO_2 emission level

Variable	World		No-OECD		OECD	
	FE	RE	FE	RE	FE	RE
	Coefficient	Coefficient	Coefficient	Coefficient	Coefficient	Coefficient
Log(income)	-5.108**	-4.906**	-1.748**	-1.836**	0.879	1.089
	(-7.26)	(-7.41)	(-7.97)	(-8.49)	(0.20)	(0.26)
Log(income)sq.	0.368**	0.355**	0.132**	0.138**	0.125	0.097
	(8.50)	(8.75)	(9.27)	(9.83)	(0.50)	(0.42)
Log(K /L)	-1.474**	-1.299**	-0.477**	-0.446**	-11.87**	-10.87**
	(-4.02)	(-3.63)	(-4.76)	(-4.55)	(-5.21)	(-5.0)
Log(K /L)sq.	0.103**	0.093**	0.038**	0.035**	0.635**	0.579**
	(4.77)	(4.41)	(5.92)	(5.62)	(5.41)	(5.15)
Log(population)	-0.0160	-0.1524**	0.069	-0.058**	0.324	-0.195
	(-0.09)	(-2.8)	(1.06)	(-3.43)	(0.53)	(-1.4)
OPEN	**6.3E-03****	**5.7E-03****	**-9.8E-05**	**-5.7E-04**	**4.5E-02****	**4.4E-02****
	(4.07)	**(3.85)**	**(-0.18)**	**(-1.14)**	**(6.78)**	**(7.5)**
Time trend	-6.0E-04	1.2E-03**	-4.3E-03*	9.2E-05	-1.6E-02	-2.2E-03
	(-0.13)	(4.21)	(-2.39)	(0.59)	(-1.47)	(-1.9)
Relative (K/L)*OP	-0.013**	-0.01**	-0.002	-0.0015	-0.03**	-0.029**
	(-8.92)	(-7.44)	(-1.2)	(-1.15)	(-8.99)	(-9.06)
Relative (K/L)^2*OP	3.8E-04	7.4E-06	-2.3E-04	-1.9E-04	2.4E-03**	2.1E-03**
	(1.33)	(0.03)	(-0.32)	(-0.27)	(4.51)	(4.16)
Relative (Income)*OP	-1.1E-03	-1.7E-03	3.8E-03**	3.7E-03**	-5.1E-03**	-5.5E-03**
	(-1.10)	(-1.83)	(5.65)	(5.57)	(-2.96)	(-3.36)
Relative (Income^2)*OP	-6.7E-05	-2.8E-05	-3.1E-04	-3.2E-04	4.2E-05	6.7E-05
	(-1.51)	(-0.65)	(-1.85)	(-1.91)	(0.59)	(1.01)
Constant		22.42**		8.14**		38.03**
		(9.76)		(9.74)		(3.09)
Hausman test		133.15**		28.74**		63.67**

Note: Figures in parentheses are t-values. * and ** denote 5% and 1% level of significance, respectively.

However, conventional trade theory suggests that capital abundant countries export the capital-intensive goods, (which is also energy intensive,) which increases pollution in capital abundant country[110]. So, it seems that the PHH and FEH contradict. Actually, the differences in environmental policy (PHH) and differences in factor endowments (FEH) might interact and jointly may determine the comparative advantage in trade. The possible interaction effects are incorporated in our last two equations i.e., eq. (8.3)-(8.4). Now we present the estimates from our equations (8.3) – (8.4) allowing for the interaction of country characteristics with a measure of openness. Tables 8.4 – 8.6 present the estimated results of the impact of comparative advantage (interaction with country characteristics) on emission level, emission intensity and relative change of emission, respectively. The results are dramatically changed. Tabless 8.4 – 8.6 present that the coefficient of openness is positive and significant for all three measures of pollution for all groups, except Non-OECD. Considering the coefficient of openness, the results of Tables 8.4 – 8.5 still support the PH effects. It should be noted that the country characteristics have made a large difference to the impact of openness on CO_2 emission. The signs of the coefficient of openness change

[110] Globalization helps to concentrate on the capital-intensive industry and thereby raising pollution.

dramatically opposing the previous results (i.e., PHH) with high significant level. This result differs from the findings of Antweiler et al., (2001), and Copeland and Taylor (2004[111]). The difference arise could be due to different pollution index used by them. They use SO_2 concentration[112] data in volume whereas we use only CO_2 emission. The interaction terms with country characteristics are also highly significant. The interaction term of openness and relative capital-labour ratio is negative and its square term is positive in most of the cases except in Non-OECD for emission intensity (see, Tables 8.4 – 8.6). We also observe the interaction terms (of relative capital intensity and openness) are negative and positive for all three measures in the world and OECD. These findings suggest that if a country has sufficiently high capital-labour ratio relative to the rest of the world, further openness makes this country clean. Relatively rich and poor country would be cleaner and dirtier[113], respectively. This implies that rich capital abundant country diverts one part of their capital and labour to produce clean and knowledge-based technology.

Table 8.5. Estimated Interaction Results on CO_2 emission intensity (per dollar)

Variable	World		No-OECD		OECD	
	FE	RE	FE	RE	FE	RE
	Estimated Coefficient	Estimated Coefficient	Estimated Coefficient	Estimated Coefficient	Estimated Coefficient	Estimated Coefficient
Log(income)	1.5×10^{-4}	2.1×10^{-4} **	1.7×10^{-4} *	1.6×10^{-4} *	1.8×10^{-3} **	1.7×10^{-3} **
	(1.91)	(2.75)	(2.43)	(2.35)	(4.07)	(4.23)
Log(income)sq.	-1.06×10^{-5} *	-1.4×10^{-5} **	-1.3×10^{-5} **	-1.26×10^{-5} **	-9.9×10^{-5} **	-9.55×10^{-5} **
	(-2.16)	(-3.07)	(-2.86)	(-2.77)	(-4.03)	(-4.18)
Log(K /L)	-2.2×10^{-4} **	-1.96×10^{-4} **	-1.03×10^{-5}	-5.55×10^{-6}	-1.6×10^{-3} **	-1.56×10^{-3} **
	(-5.21)	(-4.8)	(-0.32)	(-0.18)	(-6.89)	(-7.27)
Log(K /L)sq.	1.7×10^{-5} **	1.54×10^{-5} **	3.17×10^{-6}	2.68×10^{-6}	8.6×10^{-5} **	8.7×10^{-5} **
	(6.87)	(6.43)	(1.54)	(1.34)	(7.44)	(7.85)
Log(population)	2.4×10^{-5}	-1.39×10^{-5} *	4.2×10^{-5} *	3.45×10^{-6}	-1.05×10^{-4}	-2.75×10^{-5} *
	(1.27)	(-2.16)	(2.02)	(0.58)	(-1.75)	(-1.97)
OPEN	**1.67**$\times 10^{-6}$ **	**1.57** $\times 10^{-6}$ **	**2.45E-07**	**1.38E-07**	**5.77** $\times 10^{-6}$ **	**5.7** $\times 10^{-6}$ **
	(9.53)	**(9.33)**	**(1.42)**	**(0.85)**	**(8.83)**	**(9.80)**
Time trend	-7.41E-07	1.88E-07**	-1.22E-06*	-3.20E-08	1.77E-06	-1.90E-07
	(-1.40)	(5.62)	(-2.12)	(-0.54)	(1.68)	(-1.65)
Relative (K/L)*OP	-2.2×10^{-6} **	-1.9×10^{-6} **	6.93E-07	7.24E-07	-4.4×10^{-6} **	-4.08×10^{-6} **
	(-13.4)	(-12.15)	(1.56)	(1.67)	(-12.97)	(-13.05)
Relative (K/L)^2*OP	1.15×10^{-7} **	7.01E-08*	-7.4×10^{-7} **	-7.3×10^{-7} **	3.7×10^{-7} **	3.4×10^{-7} **
	(3.58)	(2.26)	(-3.15)	(-3.18)	(7.01)	(6.8)
Relative (Income)*OP	-1.15×10^{-7}	-2.02×10^{-7}	2.45E-08	-2.02E-09	-6.3×10^{-7} **	-6.18×10^{-7} **
	(-1.04)	(-1.88)	(0.11)	(-0.01)	(-3.7)	(-3.81)
Relative (Income^2)*OP	-1.2×10^{-8} *	-8.3×10^{-9}	5.8×10^{-8}	5.38×10^{-8}	4.35×10^{-9}	5.75×10^{-9}
	(-2.42)	(-1.69)	(1.06)	(0.99)	(0.62)	(0.87)
Constant		-1.69×10^{-5}		-5.4×10^{-4} *		-4.24×10^{-4}
		(-0.06)		(-2.02)		(-0.35)
Hausman test		116.97**		12.4		72.55**

Note: Figures in parentheses are t-values. * and ** denote 5% and 1% level of significance, respectively.

[111] Their estimated effect is quite small indicating that the FEH and PHH counteract and potentially tend to offset each other, but negative sign remain same and significant.

[112] Special econometric data analysis is more appropriate for concentration data because of transboundary problem which is not applicable to emission data.

[113] Thus, these results also partially support the PHH.

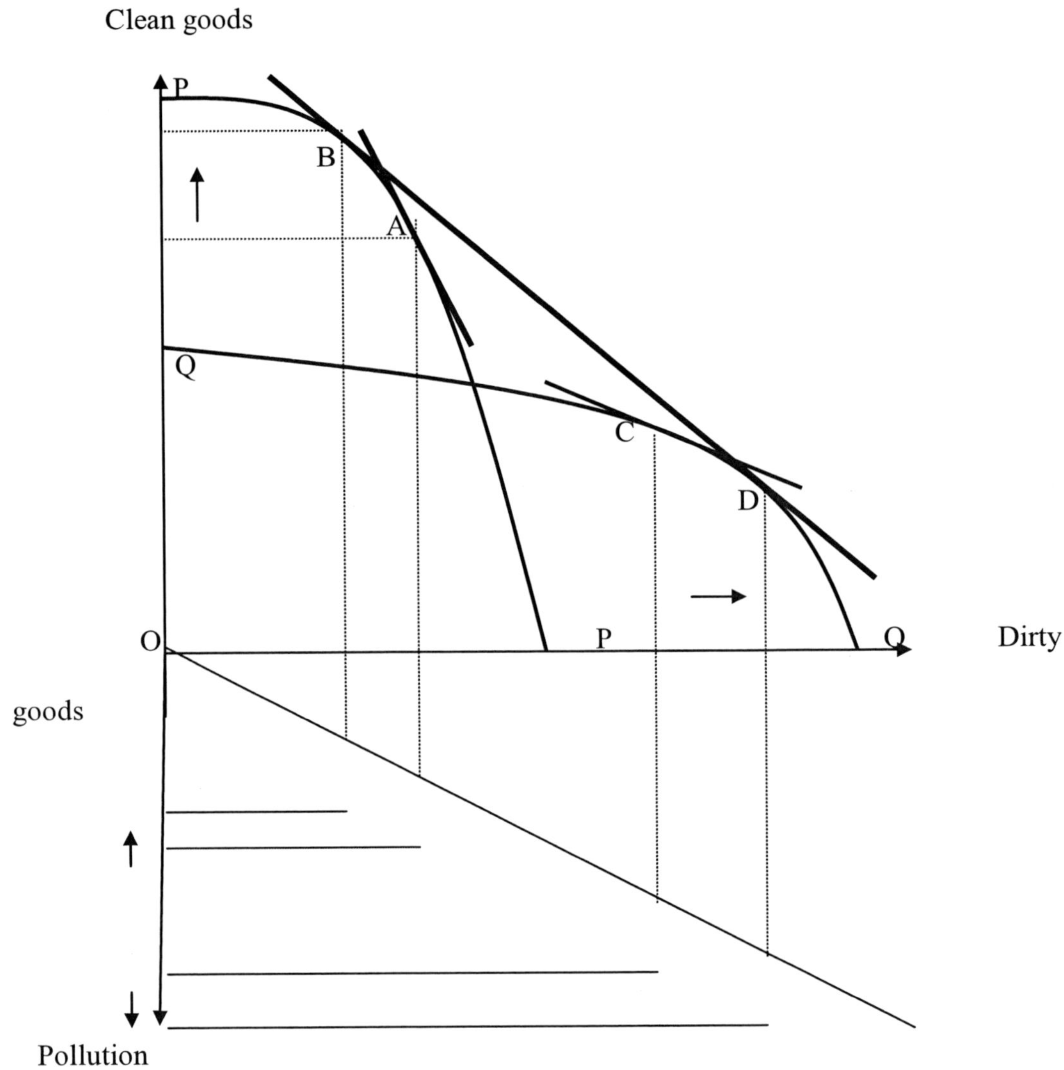

Figure 8.1. Pollution Haven Hypothesis.

The results of interaction on openness and relative income are also similar to that of capital-labour ratio. Globalization reduces emission marginally in OECD and increases in Non-OECD countries while it raises emission in the world. These empirical findings suggest that globalization increases the pollution, particularly CO_2 emission, and thereby it actually promote the global warming. It is true that net impact of globalization on emission depends on the relative strength of the factor endowments and policy regulations.

The coefficients of interaction terms in general are significantly negative but squared interaction terms are significantly positive. This implies that globalization (or openness) improves the environmental quality in (relatively) more capital intensive and economically rich countries compared to less capital and poor countries. Thus, developed nations are gaining more and more in terms of better environment. It should be noted that comparatively poor countries tend to increase emission (either in level or in intensity). This is possible only through raising capital intensity, which is the result of globalization. In other words, globalization might help to move capital from developed countries to less developed countries (i.e., capital flight) and thereby capital labour ratio increases in under developed countries. As a result there is no clear evidence of environmental better off or worse off particularly in Non-OECD countries. To explore more detail we further split the world into seven country-groups on the basis of geo-political boundaries and continue our analysis. Table 8.7 provides the results of different country-groups. Table 8.8 shows the partial effects on rich and poor after controlling other variables. The basic results or findings remain same.

Table 8.6. Estimated Interaction Results on relative change of CO_2 emission

	World		Non-OECD		OECD	
	FE	RE	FE	RE	FE	RE
Variable	Estimated Coefficient	Estimated Coefficient	Estimated Coefficient	Estimated Coefficient	Estimated Coefficient	Estimated Coefficient
Log(income)	1.6156**	2.1244**	1.868**	1.86**	6.961**	6.883**
	(5.15)	(7.15)	(4.15)	(4.16)	(5.43)	(5.74)
Log(income)sq.	-0.0447*	-0.0777**	-0.069*	-0.0695*	-0.323**	-0.321**
	(-2.31)	(-4.27)	(-2.37)	(-2.40)	(-4.50)	(-4.84)
Log(K /L)	0.0142	0.1188	-0.064	0.057	-2.874**	-2.456**
	(0.09)	(0.74)	(-0.31)	(0.28)	(-4.38)	(-3.93)
Log(K /L)sq.	0.0184	7.89E-03	0.0268*	0.015	0.156**	0.133**
	(1.90)	(0.84)	(2.04)	(1.21)	(4.61)	(4.12)
Log(population)	0.3627**	7.57E-03	0.526**	-0.014	0.399*	0.0207
	(4.83)	(0.28)	(3.96)	(-0.36)	(2.29)	(0.52)
OPEN	**6.49E-03****	**5.63E-03****	**5.52E-03****	**3.60E-03****	**0.015****	**0.015****
	(9.44)	**(8.44)**	**(5.01)**	**(3.48)**	**(7.73)**	**(9.05)**
Time trend	-0.0111**	8.36E-04**	-0.018**	-1.31E-04	-5.94E-03	-7.68E-04*
	(-5.33)	(5.43)	(-4.79)	(-0.32)	(-1.93)	(-2.37)
Relative (K/L)*OP	-5.92E-03**	-5.32E-03**	-4.77E-03	-2.99E-03	-9.73E-03**	-9.70E-03**
	(-9.22)	(-8.39)	(-1.68)	(-1.08)	(-9.99)	(-10.68)
Relative (K/L)^2*OP	6.23E-04**	4.28E-04**	2.88E-04	-2.59E-04	1.08E-03**	1.06E-03**
	(4.96)	(3.49)	(0.19)	(-0.18)	(7.02)	(7.21)
Relative (Income)*OP	-1.62E-03**	-1.88E-03**	8.21E-04	5.53E-04	-2.54E-03**	-2.72E-03**
	(-3.71)	(-4.43)	(0.60)	(0.41)	(-5.11)	(-5.77)
Relative (Income^2)*OP	2.69E-05	3.12E-05	3.24E-04	2.51E-04	5.62E-05**	6.64E-05**
	(1.36)	(1.61)	(0.93)	(0.73)	(2.77)	(3.45)
Constant		-14.98**		-12.98**		-23.97**
		(-14.48)		(-7.51)		(-6.77)
Hausman test		81.463**		37.766**		52.641**

Note: Figures in parentheses are t-values. * and ** denote 5% and 1% level of significance, respectively.

Table 8.7. Impact of openness in different continents/country-groups

Continents/ Country-groups	Without comparative advantage in trade		With comparative advantage in trade	
	Fixed Effect	Random Effect	Fixed Effect	Random Effect
	Coefficient of openness	Coefficient of openness	Coefficient of openness	Coefficient of openness
Africa	0.007** (3.95)	0.0056** (3.24)	0.036** (6.16)	0.0427** (8.1)
Central America	0.0066** (6.35)	0.0063** (6.21)	-0.0042 (-1.11)	-0.0073* (-2.15)
South America	0.0013 1.08	0.0013 (1.10)	0.0095** (4.95)	0.0082** (4.98)
America(CA+SA)	0.004** (4.96)	0.003** (4.68)	0.0072** (4.26)	0.00396* (2.55)
Asia (ex Jap)	-0.0013 -1.45	-0.0022* (-2.54)	0.006* (2.35)	0.0043 (1.82)
East Europe	0.0096** (5.34)	0.0084** (5.46)	0.0336** (8.40)	0.0315** (8.80)
West Europe	-0.0042** (-3.69)	-0.005** (-5.32)	0.014** (5.33)	0.0191** (8.32)
Europe	-0.0018 -1.70	-0.0025** (-2.91)	0.0176** (9.56)	0.0186** (11.51)

Note: Figures in parentheses are t-values. * and ** denote 5% and 1% level of significance, respectively.

Table 8.8. Conditional or Partial impact of openness on environment

	Emission Level (per capita)		Emision intensity (per $)	
	OECD	Non-OECD	OECD	Non-OECD
R I*OP	-0.0026**	0.0058**	$-1.7x\ 10^{-6}$ *	$2.3x10^{-6}$
RKL*OP	-0.0053**	0.0047**	$-1.5x\ 10^{-5}$ **	$2.9x10^{-6}$ *

Note: Figures in parentheses are t-values. * and ** denote 5% and 1% level of significance, respectively.

8.5. CONCLUSION

The aim of this chapter is to explore whether globalization helps or hurts the environment. Using panel data technique we examine the impacts of globalization on pollution level, pollution intensity and relative change of pollution for the developed (OECD) and developing (Non-OECD) country groups and the world as a whole. This study examines the factor endowment and pollution haven hypotheses that predict how international trade affects the environment. In this study we use CO2 emission, which is the mail culprit of the global warming, and observe that the impact of globalization on environment heavily depends on the basic characteristics of a country and its dominating comparative advantage. The empirical results suggest that globalization help developed countries to reduce CO_2 emission while developing countries to raise it. Net impact of globalization might increase the global worming. In this context more rigorous study is needed. Possibly in a dynamic framework we could find the true pollution reduction mechanism in liberal open economy.

Chapter 9

ENVIRONMENT AND TECHNOLOGICAL PROGRESS

9.1. INTRODUCTION

During last two decades the impact of economic growth on environment has become a focus of attention for academics and policy makers. Researchers mostly observe a non-linear or inverted-U relationship between environmental degradation and economic growth which is termed as *Environmental Kuznets Curve*[114] (*EKC*). With industrialization environmental pressure rises initially and later declines mainly due to income and technological effects. It is true that economic growth, itself paves the way for technological innovation, which improves productivity and quality of life, on other hand, it creates few problems in terms of externalities (for example, emission, waste generation etc.) that may damage the existing environment. So, economic growth improves quality of life in terms of produced goods but reduces welfare (directly or indirectly) through degradation of environmental quality. Lots of papers in the literature have discussed several issues to justify the EKC, but few have attempted to explain the EKC in the light of technological advancement in long run[115].

Smulders et al., (2005) provide an analytical foundation for EKC explaining by policy-induced technology. Following Smulders and Bretschger (2000), Smulders et al., (2005) this chapter argues how human society as a whole creates the condition for technological innovations and reduces pollution externalities[116]. The socio-economic system creates the condition for technological change. People forms environmental lobby (like NGO (Non-Governmental Organization)) and create pressure on political lobbies and/or government to change or amend laws to control pollutions. The regulatory institutions become more effective to restrict pollution externalities. So, industrialists face higher cost of production under new regulations. They must search for alternative technology (through R & D) such

[114] In 1990s, the worldwide debate starts to find the relationship between economic growth and environmental degradation. Litareture (Grossman and Krueger (1991, 1995), Selden and Song (1994, 1995), Stern (1998, 2002, 2004), Dinda et al., (2000), Dinda (2004, 2005), Dinda and Coondoo (2006), Coondoo and Dinda (2002, 2008), Andreoni and Levinson (2001), McConnell (1997), Munasinghe (1999), Bertinelli and Strobl (2005), Chimeli and Braden, (2005), Cassou and Hamilton (2004), Khanna (2002), Liu (2005), Maddison (2006), Smulders and Bretschger (2000), Smulders et al., (2005) etc.) assert that environmental quality deteriorates in early stage of economic development and later improves.

[115] Quality of life changes with technological innovations.

[116] In other words, technological progress might improve the efficiency either in terms of productivity or in terms of fuel use per unit of output. It also drives towards efficient use of input resources.

that production is efficient in terms of cost as well as externalities[117]. Thus, the society as a whole creates the condition to innovate and adopt new technology, which definitely reduces pollution level[118]. Thus, as income rises, environmental regulation is tightened that spurs pollution reducing innovation, which is consistent with Porter hypothesis (see Mohr (2002)) also. As a result, upgraded new and cleaner technologies replace the dirty and obsolete technologies[119]. Truly, continuously society creates the condition for innovation and thus, we observe a web of technological innovations in very long run. This paper tries to explore this untouched part and attempt to explain the possible EKC envelope with technological progress in very long run.

Purpose of this chapter is to explain EKC in terms of technological development in long run. Specifically, how human society creates condition for innovations towards green technology or shifting of fuel based technology such that pollution externality declines and as a consequence EKC changes. Intuitively we perceive a possible envelope of EKCs that are related to specific technologies. The study follows as: Section 2 discusses the motivation of this study. Section 3 develops a simple economic growth model and finally concludes.

9.2. Motivation

Innovation first develops useful new products or/and production processes that improves quality of products and/or processes. Innovation solves problems of scarcity by creating new materials, which may solve problems of earlier technology by developing clean products. Obviously, new product is better than previous one. Gradually people come forward to adopt the new technology and soon it becomes a mass adopted technology prevailing in the economy[120]. However, after sometime, it turns out that the new products, processes and materials have their own problems[121], which are unanticipated at the time of installation. New technologies, unambiguously, improve productivity but may create potential dangers to society such as new hazardous wastes, risk and other human health related problems (direct or indirect). These problems trigger public responses through socio-political and economic process and call for action regarding laws and regulations. If necessary, new social institutions are set up for monitoring, control and regulations. All these stimulate to search for another technology which may solve the problems, but these new innovations may create again another unknown new problems or externalities. Thus, in long run, society or economy

---

[117] Tarui and Polasky (2005) developed a model of environmental regulation with learning about environmental damage and endogenous choice of abatement technology.

[118] Reis (2001) study the effect on optimal growth of the possibility that at some moment in future a technology is discovered that eliminate pollution.

[119] See, Smulders, Bretschger and Egli 2005 for details. They provide empirical evidence for few European countries (UK, Germany and Switzerland) and USA.

[120] See also the so called general adopted technology (GAT) (Smulders and Bretschger 2000, Bresnahan and Trajtenberg 1995).

[121] With regard to the technique effect there are many examples of efficient resource use, substitutions between resources, and containment of wastes. The most dramatic are reduction of SO_2 in Germany, France and Japan by the installation of flue-gas desulphurization equipment (in Germany), a switch over to nuclear power (in France), and a combination of these two (in Japan). However, both alternatives have secondary environmental effects, i.e., quarrying and transport of large quantities of limestone for flue-gas desulphurization, waste disposal, radioactive emissions and risk of accidents for nuclear power (For example, accident at Chernobyl in Russia). After adoption of said technologies these effects are identified.

as a whole may produce series of technologies. Few examples of broad technological events are given below briefly.

Broadly technology has gradually shifted from animal power to fossil fuel to nuclear power[122]. In the second half of the twentieth century, nuclear power held great promises at the time of introduction but resistance against it grew rapidly. The mad-cow disease[123] exemplifies how new technology already has created new health problems. Recently, genetically modified crops may improve productivity in agriculture, but at the same time public become aware and increasingly worried about possible side effects, even though scientific research has not yet claimed or supported these worries. These examples have a cyclical pattern in common.

A cyclical pattern arises in technologies, which first diffuse, then become regulated and finally are phased out by next generation of technology (Smulders and Bretschger 2000, Smulders et al., 2005). Initially pollution externality may increase with rapid adoption of a technology, which also generates income of the society. After mass adoption of the technology some problems or externalities may come out. Then society raises voice against the (bad) externalities and social forces take actions (like enforcement) to stop it. Truly, social forces are prime mover and direct scientific efforts towards new directions. Thus, scientific affords might be concentrated on new directions of innovations[124] to solve the social problems. Thus, society as a whole stimulates to generate new technology. Pollution level of old technology declines as the existing old technology is replaced by new one (without hampering economic activities[125]). Thus, an inverted-U shaped or *EKC* is observed for each technology. Since technology itself changes over time, in long run, a series of technologies emerge and correspondingly we may observe series of EKCs and all these might produce an envelope that may take shape of nonlinear relation such as inverted-U, J or *N*.

9.3. MODEL

9.3.1. Welfare

Consider an economy consist of single economic agent who operates all the economic activities. This representative household maximizes her welfare or obtains utility consuming

[122] Steam-powered cars and coaches replaced bull or horse carts and freed the inner cities of the nuisance of animal manure, which was a serious problem in the nineteenth century (Smulder and Bretschger 2000). New technology of the nineteenth century (viz., the innovation of the steam engine (1765)), steam-powered cars, created another problem of huge coal dust, flying ash and CO_2 emissions. Fuel-powered (petrol, diesel) cars and coaches replace steam-powered cars in twentieth century, and reduce coal dust, flying ash and CO_2 emissions but has created other pollutants like hydrocarbon, lead (Pb), SO_2, NO_x, CO_2 and carbon monoxide (CO) emissions etc. They have created problems in the form of air pollution in the cities and also a serious problem of global warming in the twenty-first century.

[123] Feeding sheep meat to cows was intended to boost milk production and later it was found out that there was a connection to Creutzfeld-Jacob.

[124] Using sensitivity analysis Popp (2004) shows that potential crowding out of other R & D and market failure in the R & D sector are the most important limiting factors to the potential of induced innovation.

[125] Each stage of economic development may also constitute a series of technological up gradation. As technology changes over time, in long run, a sequence of EKCs might emerge. Thus, we may also have a smooth envelope of series of new generations of upgraded technology.

goods c and dis-utility (lose of welfare) for pollution, p, (denotes the pollution index), which is generated in production process of c goods. The utility function of the household is

$$U(c,p)\ \ U_c > 0, U_{cc} < 0,\ U_p < 0, U_{pp} > 0,\ U_{cp} < 0 \tag{9.1}$$

In the infinite time horizon inter-temporal choice problem, the representative agent maximizes his/her present value of utility (or welfare of the society):

$$\text{Maximize } W = \int_0^{\infty} e^{-\rho t} U(c,p)\, dt \tag{9.2}$$

Where ρ (>0) is the discount rate.

9.3.2. Production

Considering a one- good closed economy, output is produced by only composite capital, k, for a given technology. Production function of this economy is

$$y = f(k),\ f_k > 0 \text{ and } f_{kk} < 0. \tag{9.3}$$

Let specific production function be

$$y = Ak^{\beta}\ \ 0 < \beta < 1 \tag{9.4}$$

A is the technology. So, the production of the economy depends only on composite capital k, which also generates pollution as a by-product, but technological improvement reduces it.

9.3.3. Pollution

Pollution is unavoidable and an inherent relation with production process using capital for any available technology. Let μ be the rate of pollution is fixed proportion of output. Pollution rate, μ, may be a decreasing function of technological improvement. For simplicity I assume constant μ. Pollution is generated directly with production but declines with available cleaner technology. More specifically, each time one technology may produce another or more than one technology. So, technology (A) produces technology (A) that can be expressed by A^2. Thus, A^2 may crudely represent technological availability that definitely helps to select low pollution generating process. This is perhaps a strong assumption in this model set up. So, the pollution flow at each moment is proportional to output production and inverse to technological availabilities, i.e.,

$$p = \frac{\mu y}{A^2},\ 0 < \mu < 1 \tag{9.5}$$

Where A can be understood as an index of technology used, and higher A (i.e., here A^2) means available cleaner technology (See, Reis 2001) in the economy.

Substituting eq.(9.4) into eq.(9.5),

$$p = \frac{\mu k^{\beta}}{A} \tag{9.6}$$

Now, stock of capital for given technology at each moment determines pollution, p (denotes the pollution index). Pollution is generated directly with production input (k) for given technology (A) at a given time, however, over time technology (A) itself changes. Thus, stock of capital and technological progress jointly determine pollution, p, in long term[126].

9.3.4. Technological Development

Technology grows exponentially at a constant rate, θ, say. Following textbook literature, technological progress is written as

$$A(t) = A_0 e^{\theta t} \tag{9.7}$$

Where A_0 is initial level of technology and technological growth rate θ (>0), which is arbitrary constant. This θ may vary with stock of capital (k) from country to country. For example, θ is small or close to zero for less developed countries with low capital but θ is high in developed countries with high stock of capital. In this context it should be mentioned that truly technological development should be a function of capital and time, i.e., $A(k,t)$. Technological progress occurs in capital - time space only (See, Basu and Weil (1998) for details). For analytical purpose I concentrate only in one dimension.

Following Mohr (2002), now, I clarify the conceptual definition of capital

$$k_t = \int_{-\infty}^{t} l_\tau d\tau \tag{9.8}$$

---

126 In general one fraction of capital is used in production process, which generates pollution and remaining part should be utilized for R & D (abatement) activity to reduce pollution for (long run) sustainable development (See, Dinda 2005).

Where l is the amount of labour devoted for work in each time for given technology (See, Mohr (2002) for details). Here, capital is simply defined as accumulated experience or acquired knowledge devoted by labour time in given technology. Truly, k_t is the accumulation of knowledge, which is the prime mover for technological innovation. Technology upgrades with stock of (knowledge) capital over time. It is a continuous process.

Amount of stock of capital is simply the indicator of the stage of economic development. One economy moves from less developed to developed stage mainly with raising stock of capital by accumulation over time. Since k grows over time, it is reasonable to replace time t by capital k, (i.e., $t = k$, as $t \to 0, k \to 0$ and $t \to \infty$, $k \to \infty$). This is simple assumption[127] about the linear relationship between them. There should be more complex non-linear relationship between k and t in the real world. For simplicity and analytical tractability, I consider very simple linear relation between k and t. So, this changes dimensions from time scale to capital scale. Now, I replace t by k, and also θ by ϕ.

$$t = k,\ \theta = \phi \tag{9.9}$$

Thus, now technology (A) itself is a function of capital i.e., $A(K)$ and substituting eq.(9.9) in eq.(9.7), it turns to

$$A(t) \equiv A(k) = A_0 e^{\phi k} \tag{9.10}$$

The eq. (9.7) and eq. (9.10) are identical but interpretation is different. Now ϕ (>0) can be interpreted as the measure of effectiveness of capital on technological innovation. The rising capital induces to improve technological innovations, which helps to reduce pollution through technological development, which allows production to generate less pollution.

Now, substituting eq.(9.10) in eq. (9.6),

$$p = Bk^{\beta} e^{-\phi k} \tag{9.11}$$

Where $B = \mu A_0^{-1}$

Pollution, p, is directly related with stock of capital at diminishing rate and inversely with technological development (measured in k-scale), which depends on ϕ and k. Thus, technological development depends on the effectiveness of capital on technological innovation that definitely reduces pollution. Thus, equation (9.11) [128] depicts a non-linear relationship between pollution (index), p, and stock of capital, k (Figure 9.1c).

[127] Here time scale is applicable only for development period during which capital accumulates.
[128] Equation (9.11) looks very similar to Beltratti (1996).

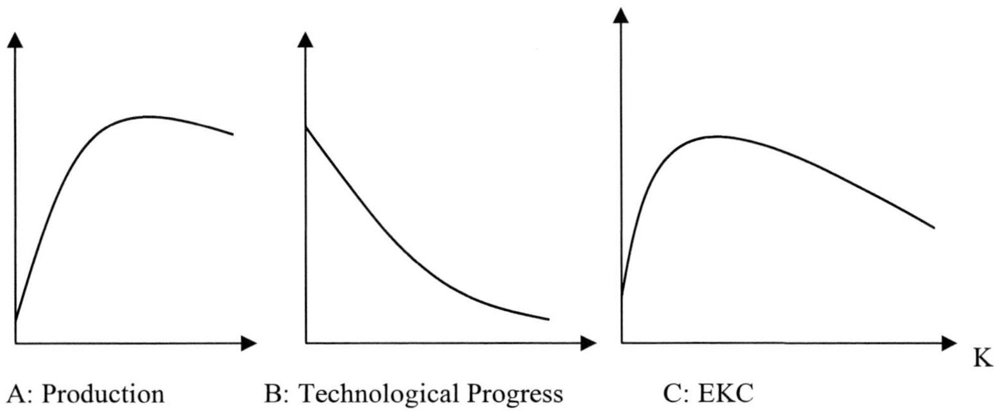

Figure 9.1. Relationship between pollution with capital and Technological Progress.

Now we examine the properties of eq. (9.11). Proportional change of pollution depends on change of capital stock (*dk*), share of capital ($d\beta$) and technological innovation sensitivity ($d\phi$).

$$\frac{dp}{p} = \left[\left(\frac{\beta}{k} - \phi\right)dk + (\ln k)d\beta - k\,d\phi\right] \tag{9.12}$$

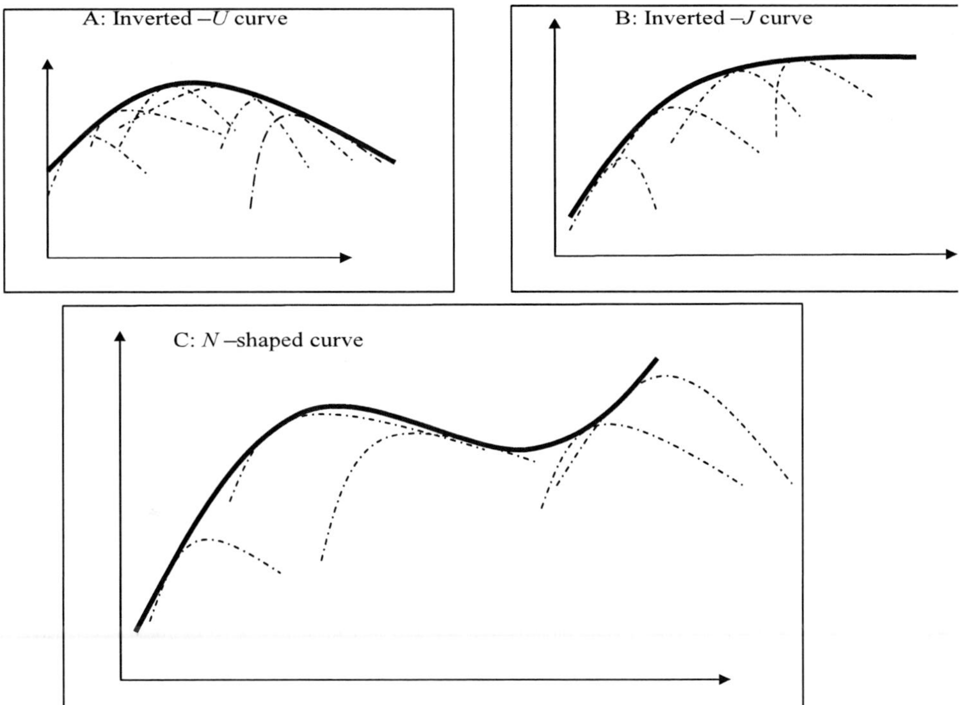

Figure 9.2. Possible shapes of the envelope of EKCs.

Thus, proportional change of pollution directly related to the change of capital and its share but inversely related to the effectiveness of technological innovation. Shape of the curve depends on the relative strength of these variables. In general, β is constant, so $d\beta = 0$, and for simplicity I also consider ϕ is fixed (later I relax this assumption).

Pollution directly related with capital at low level of capital stock (as long $\beta/k > \phi$, as $k \to 0$ $\beta/k \to \infty > \phi$) and inverse relation at high level of capital (as $k \to \infty$ $\beta/k \to 0 < \phi$, $\beta/k < \phi$). Intuitively it is clear that if one economy moves from low level of capital to high, then its pollution generating path will be inverted-U shaped. Initially p rises then reaches at maximum, thereafter declines. Intuitively the high stock of capital stimulates to invent new technology which phases out old, and consequently pollution declines. Thus, pollution declines with technological development at high level of k. For each time there is one *EKC* or inverted-U. Over time there will be several *EKC*. These series of *EKC* curves should produce also a long run relationship between pollution (p) and capital (k). This long run relationship may take any shape like inverted-U, inverted-J or N-shaped curve that are generated from the envelope of *EKCs* related to corresponding technologies (Figure 9.2).

9.3.4.1. Case 1: Only k Changes

It should be noted that equ.(9.11) depicts the inverted-U or EKC relationship between p and k, given β and ϕ. For a given technology (assuming $d\phi = d\beta = 0$), pollution is maximum at k^*, ($\frac{dp}{dk} = 0$ at $k^* = \frac{\beta}{\phi}$), increases (i.e., $\frac{dp}{dk} > 0$) as long as $k < k^*$, and declines (i.e., $\frac{dp}{dk} < 0$) when $k > k^*$. The turning point is observed at $k^* = \frac{\beta}{\phi}$. It should be noted that the output elasticity (β) and sensitivity of technological innovation (ϕ) jointly determine the so-called turning point. For given value of β and ϕ, the value of turning point k^* remains unchanged, but only the peakness of *EKC* differs due to variation of initial level of technology, i.e., A_0 (Figure 9.A.1).

The effectiveness of technological innovation (ϕ) has positive impact on the economy but negative effect to pollution externality. Turning point K^* may change as ϕ changes with accumulation of capital and technological development or/and change of output elasticity β. It should be mentioned that pollution features also varies with type of power[129] used in economic activities as mentioned in earlier section.

9.3.4.2. Case 2: Only ϕ Changes

Effectiveness of innovation sensitivity actually depends on the stock of (knowledge) capital which is directly related to the stage of economic development. It is clear from

[129] In general it is approximately constant. The output elasticity may change only when production technique changes that depend on shifting the use of (fuel) power like animal (horse) power to steam (coal) power, steam to diesel, diesel to electric, and electric to nuclear power, etc.

equation (9.10) that as $\phi \to 0$, $p \to Bk^{\beta}$; i.e., pollution is ever rising, the turning point ($k^{*} \to \infty$) approaches to infinity. For given β, as ϕ decreases turning point k^{*} increases (See Figure 9.A.2, generated by simulation). This suggests that if technological innovation sensitivity is ineffective then pollution monotonically increases (for example, CO_2 emission) while reverse case in high innovation sensitivity, i.e., as $\phi \to \infty$, $p \to 0$, turning point k^{*} does not exist. This implies that if technological innovation is highly sensitive then pollution comes down to zero (negligible) level.

Proposition 9.1: *Sensitivity of technological innovation generates the shape of pollution curve.*

Now rewritten the equation (8.12) as

$$\frac{dp}{d\phi} = \left[\left(\frac{\beta}{k} - \phi\right)\frac{dk}{d\phi} + \ln k \frac{d\beta}{d\phi} - k\right]p \tag{9.13}$$

Since $\frac{dk}{d\phi} > 0$, for given β (i.e., $\frac{d\beta}{d\phi} = 0$), pollution may rise or fall that depends on relative strength of (-*k*) and $\left(\frac{\beta}{k} - \phi\right)$. Over all pollution level may increase, $\frac{dp}{d\phi} > 0$, as long as stock of capital is less than desired stock of capital with $\frac{dk}{d\phi} > 0$, and p may decline with effectiveness of capital on technological innovation, $\frac{dp}{d\phi} < 0$, when stock of capital exceed desired capital level. Pollution may be maximized at $\frac{dp}{d\phi} = 0$ only when $\frac{dk}{d\phi} = \frac{k^2}{\beta - \phi k}$, provided $\beta \neq \phi k$. In this context, the sensitivity of innovation can be analyzed in terms of elasticity, i.e., $\eta = \frac{dk}{d\phi}\frac{\phi}{k} \equiv \frac{\phi k}{\beta - \phi k}$. As long as $\beta > \phi k$ then innovation sensitivity is positive and highly sensitive or (infinitely) elastic when pollution level reaches at peak ($\beta = \phi k$). Economy moves from one technology to another when $\eta < 0$ which is corresponding to $k^{*} < k$. In other word, old technology is replaced by new one. Actually, η provides the signal for switching over from one technology to another. Society or economy allows to raise pollution with existing technology until $\eta < \infty$ but at $\eta = \infty$ economy desperately search for alternative technology and shifts to alternative technology then η turns to be negative (i.e., $\eta < 0$) that implies $\frac{dp}{d\phi} < 0$ which corresponding to $k^{*} < k$.

Proposition 9.2: $\eta < 0$ *provides definite replacement of polluting technology by an alternative one.*

In case of unitary elasticity $\eta = 1$, $\phi k = \frac{1}{2}\beta \Rightarrow k = \frac{1}{2}\frac{\beta}{\phi} \equiv \frac{1}{2}k^*$, in case of elastic, $\eta > 1$, $\phi k > \frac{1}{2}\beta$, $\eta = \infty$ at $\phi k = \beta$, and inelastic when $\eta < 1$, $\phi k < \frac{1}{2}\beta$.

Remarks 9.1: *Innovation sensitivity varies with economic development or capital accumulation.*

'Tunneling through' potential EKC (Munasinghe 1999) to avoid relatively higher level of environmental harm is possible only when $1 < \eta < \infty$, and when $\eta = \infty$ at $k = k^*$, the economy desperately search for alternative technology and immediately replaces the old technology by new one. In reality, each ϕ produces one EKC. The parameter ϕ changes over time. So, over time series of EKC is generated by series of ϕ. It may produce long run EKC which might be an envelope.

9.3.5. Pollution Sensitivity and Welfare

Representative household is sensible to pollution externality that badly affects his/her welfare. So, representative agent maximizes her welfare consuming *c* goods with minimum dis-utility for pollution. This (lose of welfare due to pollution) stimulate to gather more information and generate ideas or knowledge to prevent it. Suppose the specific utility function of household is

$$U(c, p) = \ln c - \upsilon \ln p \qquad (9.14)$$

Where $\upsilon \geq 0$. Sensitivity of pollution depends on knowledge or information that is generated by R & D, which also depends on socio-economic conditions. In less developed economy, poor people are more interested in job and income than clean air and water (Dasgupta et al., 2002). They are too poor to pay for abatement (R & D), or disregard environmental consequences of economic growth. After a considerable growth, as income rises people value the environment more (McConnel 1997, Khanna 2002) and nation can afford to spend more on R & D (Komen et al., 1997). In general, people of Less Developed Countries (LDCs) are not much aware about the effect of pollution or its consequences, and so their pollution sensitivity is very low (i.e., $\upsilon \approx 0$); whereas in Developed Countries (DCs) high level of R & D diffuse the knowledge and people become more sensitive with high value of υ (i.e., $\upsilon > 0$). Welfare reduces only if sensitivity of pollution is sufficiently high ($\upsilon > 0$). Substituting eq.(9.11) into eq.(9.14), we get

$$U(c, p) = \ln c - \upsilon \ln B - \upsilon\beta \ln k + \upsilon\phi k \qquad (9.15)$$

Suppose $B = 1$, then eq.(9.15) becomes

$$U(c,p) = \ln c - \upsilon\beta \ln k + \upsilon\phi k \tag{9.16}$$

From eq.(9.16), stocks of capital reduce welfare (through $\upsilon\beta$) and simultaneously stimulate innovation that definitely raises welfare (by $\upsilon\phi$) which provides incentive to innovate or upgrade technology.

Proposition 9.3: *Joint interaction of pollution sensitivity (υ) and effectiveness of technological innovation (ϕ) improves welfare of the society, given k.*

Pollution sensitivity can create the condition for technological innovation impulse for given stock of capital. So, for given high capital stock, ϕk can be effective only when υ is able to sensitize the condition for technological improvement. Consider an economy with low level of capital, technological innovation is insensitive, i.e., $\phi \to 0$ and welfare (utility) of this economy decreases with capital accumulation because poor people are insensitive to pollution or unaware about their loss of welfare. Whereas technological innovation is highly sensitive ($\phi \to \infty$) in developed economy with higher level of capital that helps to improve or stop further loss of welfare. Welfare reduces only if pollution is insensitive and technological progress is insufficient.

9.3.6. Economic Growth

Now, planner of the society maximizes welfare equation (9.16) subject to capital constraint

$$\dot{k} = Ak^{\beta} - c - \delta k \tag{9.17}$$

The economic growth path is

$$\gamma_c = \frac{\dot{c}}{c} = A\beta k^{\beta-1} - (\rho + \delta) + \upsilon\left(\phi - \frac{\beta}{k}\right)c \tag{9.18}$$

Economic growth crucially depends on pollution sensitivity (υ), technological innovation sensitivity (ϕ) and output elasticity - capital ratio, i.e., $\frac{\beta}{k}$. As long as $k < \frac{\beta}{\phi}$ or $k > \frac{\beta}{\phi}$, growth rate in equation (18) is lower or higher compared to pollution free world (i.e., $\upsilon = 0$, or standard Solow type growth model), respectively. Society realizes that pollution reduces welfare as long as $k < \frac{\beta}{\phi}$ and improves it when $k > \frac{\beta}{\phi}$. Individuals or society provide some affords for innovating better technology that generates more welfare and less pollution, that implies, ϕ tends to rise. Thus, individuals or society has definite positive incentive to innovate or upgrade technology for betterment of the society.

9.4. CONCLUSION

Technological innovation improves productivity and thereby quality of life, simultaneously it also creates pollution externalities that reduce welfare of the society. Social institutions formulate laws and regulations that warrant harmful externalities and finally phased out by new technology. Each technology creates new externalities, which again follow EKC. Pollution sensitivity and effectiveness of technological innovation jointly determine the shape of EKC. Truly, a cyclical pattern arises in all technologies and an envelope of them may exist in very long run.

Chapter 10

FUTURE DIRECTION OF RESEARCH

The earlier chapters of this book together have provided a somewhat detailed discussion on the problem of linkage of environmental quality with economic growth/development. This chapter summarises the main findings, discusses some limitations and indicates potential directions of future research.

This book provides a theoretical analysis in endogenous growth framework and explains the possible relationship between environment and development. This book attempts to find out the relationship and causal mechanism through which economic growth links to environmental degradation. It applies some new econometric methods in the panel data set up and illustrates the use of these new analytical tools. The empirical analyses presented in this book are all based on cross-country panel data sets. The pattern of cross-country income distribution has significant effect on the mean level of emission for all the country-groups considered. The results for the developed country-groups are found to be qualitatively different from those of their poorer counterparts.

While these findings may be interesting, the limitations of the present study should be borne in mind. First, the empirical research reported here has mostly been carried out for only one pollutant, viz., CO_2 emission mainly for an illustrative purpose. To establish our findings more conclusively, further evidence for other pollutants is required. Second, many other potentially important questions connected with causality between economic growth and environmental degradation have not been examined simply due to severe data constraints. A comprehensive analysis of income-emission relationship would necessarily call for examination of the effects of such determinants of fuel used as the sectoral composition of income, the openness of the economy and the price of fuel, among other things. In this book, we tried to examine the effects of these possible determinants of income-emission relation only tangentially. One should take into account the above mentioned factors while analyzing the income-emission relationship using a much broader framework of study.

We end this book by pointing out some other potentially important research areas connected with this study. We have already mentioned these at the concluding section of the relevant chapters. Here we shall briefly summarize the possible extensions.

For any meaningful policy discussion about control of global CO_2 emission one should require a careful examination of the nature of causality that is operative in individual countries. The concentration curve analysis has been designed at the level of a country-group, both the mean emission level and the cross-country disparity of emission level is likely to be significantly affected by the size of international trade that individual countries are engaged in. In fact, the so-called *Pollution Haven Hypothesis* asserts that developed countries are likely to import pollution intensive goods from poorer developing countries and are thereby shifting their emission to later countries. When this happens, the cross-country disparity reflected in the *SCC* and *SCR* may not be as closely linked up to the cross-country disparity of income reflected by *LC* or *LR* of income. In this context, one should bring in an appropriate measure of *openness* of the individual countries in to the analysis to identify the pure partial effect of income distribution on the emission level. The results are completely different under comparative advantage in globalized world. The whole human society depends on the technological progress in long run and overcome all problems in future.

Apart from all these, socio-political conditions should play a crucial role for long run sustainable development. Social institution, enforcement and fare governance might help to improve environmental quality.

Consider energy supplied by developing economies and that was demanded by developed economies in the framework of North-South trade. This demand- supply analysis in real terms can provide the direction of future research.

These, once carried out, will undoubtedly throw new lights on the linkage of economic development/growth with environmental degradation. However, we leave all these as the agenda for future research.

APPENDIX A

Table A.1. Continent-wise list of country groups and countries covered

Continent	Country Group	Countries Covered
Africa	Africa	Algeria, Cameroon, Cape Verde Island, Central African Republic, Comoros, Congo, Egypt, Gabon, Gambia, Ghana, Guinea, Guinea Bissau, Kenya, Madagascar, Mali, Mauritania, Mauritius, Morocco, Mozambique, Nigeria, Senegal, South Africa, Togo, Tunisia, Uganda, Zimbabwe.
America	North America	Canada and USA
	Central America	Costa Rica, Dominican Republic, El Salvador, Guatemala, Honduras, Jamaica, Mexico, Nicaragua, Panama, Trinidad & Tobago.
	South America	Argentina, Bolivia, Brazil, Chile, Colombia, Ecuador, Paraguay, Peru, Uruguay, Venezuela.
Asia	Japan	Japan.
	Asia (excluding Japan)	China, Hong Kong, India, Indonesia, Iran, Israel, Jordan, Korea Republic, Philippines, Singapore, Sri Lank, Syria, Thailand.
Europe	East Europe	Austria, Czechoslovakia, Finland, Greece, Turkey, Yugoslavia.
	Western Europe	Belgium, Cyprus, Denmark, France, West Germany, Iceland, Ireland, Italy, Luxembourg, Netherlands, Norway, Portugal, Spain, Sweden, Switzerland, U.K.
Oceania	Oceania	Australia, Fiji, New Zealand, Papua Guinea.

APPENDIX B

Econometric Methods Used

A: Introduction

In this dissertation we have used two different econometric methodologies, which are described briefly in this appendix. Section B discusses the econometric methodology of Granger Causality Test (GCT) and Section C fully describes the procedures of panel unit root test and cointegration and related error correction model (ECM).

B: The Methodology of GCT Used

In this Appendix we describe the methodology of GCT that we have used in the present exercise (Granger (1969), Hamilton (1994)). As already mentioned, the GCT is a statistical technique that helps detect the nature of causality between two variables X_t and Y_t that may be present in a given time series data set on the variables. Application of the test requires the time series of the concerned variables to be stationary. Thus, it is necessary to examine first whether the time series of the variables are stationary or not. In case they are found non-stationary, they are transformed into stationary series by successive differencing until the differenced series become stationary. The augmented Dickey-Fuller *(ADF)* and Phillips–Perron *(PP)* unit root tests are used to test for presence of unit root in the original/differenced time series of the variables to ascertain the required stationarity[130]. The *GCT* for a pair of non-stationary (integrated of order 1) variables X and Y is performed by estimating the following pair of *autoregressive distributed lag* regression equations:

$$\Delta Y_t = \alpha_{11} + \sum_{i=1}^{T_{11}} \beta_{11i} \Delta Y_{t-i} + \sum_{j=1}^{T_{12}} \beta_{12j} \Delta X_{t-j} + u_{12t}, \tag{B.1}$$

$$\Delta X_t = \alpha_{21} + \sum_{i=1}^{T_{21}} \beta_{21i} \Delta X_{t-i} + \sum_{j=1}^{T_{22}} \beta_{22j} \Delta Y_{t-j} + u_{22t}. \tag{B.2}$$

---

130 In the present exercise, the country-level time series data on *PCGDP* and/or *PCCO2* were mostly found to be integrated of order 1, rather than being stationary. The original time series of *PCGDP/PCCO2* were observed to be stationary, however, that number of such cases is very few.

Here Δ denotes the difference operator, Y_t and X_t denote the values of the variables *X* and *Y* at time *t*, respectively, *T's* denote the number of lags, α's and β's denote regression parameters, and u_t's denote the equation disturbances. The null hypothesis that X_t *does not cause* Y_t is rejected, if H_0: $\beta_{12j} = 0$, for all *j* is rejected. Analogously, rejection of the null hypotheses H_0: $\beta_{22j} = 0$ for all *j* signifies that Y_t does not cause X_t. F tests are performed to test these hypotheses.

The *GCT* gives rise to one of the following four conclusions: (i) *X* causes *Y*, but *Y* does not cause *X*, (ii) *Y* causes *X*, but *X* does not cause *Y*, (iii) both *X* and *Y* cause each other, and finally (iv) neither *X* causes *Y* nor *Y* causes *X.* While (i) and (ii) are cases of unidirectional causality, (iii) and (iv) correspond to bidirectional causality and absence of causality, respectively (see, Hamilton (1994) for details).

In the present exercise the *GCT* as described above was used to examine income-emission causality. To be precise, we have taken $X_t = \log(PCGDP_t)$ and $Y_t = \log(PCCO2_t)$, so that $\Delta X_t = \Delta \log(PCGDP_t) = r_t$ and $\Delta Y_t = \Delta \log(PCCO2_t) = r_t^*$, r_t and r_t^* measuring the growth rate of income and emission, respectively. The test was performed at the level of country groups/continents in several ways. One set of results was obtained by applying this test to the country group/continent level aggregate time series data (obtained by aggregating the country level data of all the countries belonging to a given country group/continent). A second set of results were obtained by applying the test to country group/continent-specific panel data set using ordinary least squares (OLS) to estimate the GCT regression equations treating them as *fixed effect* specifications. Since OLS may not yield efficient estimates for a panel data set, we obtained a third set of results using the *within*[131] regression method to estimate the GCT regression equations mentioned above. While estimating equations (B.1)-(B.2), we considered up to two lagged values of the variables as regressors. In most of the cases the optimal lag length (i.e., the optimal number of lagged income and/or emission variables taken as regressors) was also found to be two. It may be mentioned here that the application of *GCT* on a panel data set is not a standard practice. However, considering the fact that use of panel data increases the degrees of freedom enormously (and hence may help give robust results), we applied the *GCT* to the panel data set (see, Maddala and Kim (1999), for a discussion on the motivation behind using panel data for unit root tests).

C: Methodology Used in Panel Unit Root Test

As already mentioned in chapter 6 of this dissertation we have examined whether income-emission data for different country groups were cointegrated using the Engle-Granger bivariate cointegration analysis framework and estimated *ECM* for country groups for which cointegration was observed to be significant, using econometric techniques appropriate for a

[131] The *within* method of estimation of a linear regression equation from panel data is a method in which the slope parameters of the regression equation are estimated by applying least squares method to the data set in which for every variable an individual *unit's* observations are measured as deviation from the corresponding *unit mean*. Thus the *within* regression is one without an intercept term. This method is more frequently used to

panel data set[132]. The econometric exercise involved three steps. In the first step, the unit roots test was performed to ascertain whether or not the time series of the variables (i.e., natural logarithm of *PCGDP* and *PCCO2*, henceforth denoted by y_t and x_t, respectively) contained stochastic trend. In the second step, cointegration of income and emission was examined. Finally, in the third step, the *ECM* was estimated for those country groups for which cointegration of income and emission had been found.

In the first step the *IPS* panel data unit root test procedure was used to test presence of unit root in the time series data sets for individual country groups. The same procedure was also used in the second step while performing the Engle-Granger bivariate cointegration analysis. Finally, the *ECM* in the third step was estimated by using panel data regression technique. In what follows, we describe briefly the econometric procedures that we have used in the three steps of the present exercise.

C.1. IPS Unit Root Test

For a balanced panel data set $(y_{it}, i = 1,2,...,N; t = 1,2,...,T)$, where *i* and *t* denote cross-sectional unit and time, respectively; Im et al., (2003) considered the following linear regression set up for developing their panel unit root test

$$y_{it} = \rho_i y_{it-1} + \sum_{j=1}^{p} \theta_j \Delta y_{it-j} + z'_{it}\gamma + \varepsilon_{it} \,. \qquad \text{(C.1)}$$

Here $z'_{it}\gamma$ denotes the deterministic component of y_{it} which may be zero, a common constant intercept, a time-invariant fixed effect μ_i or a fixed effect that varies both across *i* and over *t* and ε_{it}'s are white noise equation disturbance terms. Note that in (C.1) the autoregressive parameter ρ_i is allowed to vary across units[133]. The null hypothesis for the *IPS* unit root test is H_0: $\rho_i = 1$ for all *i* and the corresponding alternative hypothesis is H_1: $\rho_i < 1$ for at least one *i*. As ρ_i is allowed to vary across *i*, the *IPS* test procedure is based on the average of the unit-specific unit root test statistics. Specifically, this test uses the average of the unit-specific *Augmented Dickey Fuller (ADF)* test statistics, which has been called the *t-bar* statistic. This is as given below:

estimate a *random effects* panel data regression model than for estimating the corresponding *fixed effect* regression model. See, Baltagi, (1999).

132 As is well known, the *ECM* is a comprehensive linear regression equation specification which provides a description of the possible nature of interdependence of the short run movements of a pair of co-integrated variable keeping in view the fact that they bear a long run equilibrium relationship.

133 Quah (1994) considered equation (C.1) without the second and third terms as the model for his panel unit root test. Levin and Lin (1993) considered a more general model to allow for fixed effects, individual deterministic trends and heterogeneous serially correlated errors. In fact, they considered equation (C.1) without the second term as their model specification. They, however, assumed the units to be *iid* $(0, \sigma_\varepsilon^2)$ and also $\rho_i = \rho$ for all i. Here H_0: $\rho = 1$ against H_1: $\rho < 1$. Levin and Lin's test is thus restrictive as it requires ρ_i to be the same for all i.

$$\bar{t} = \frac{1}{N}\sum_{i=1}^{N} t_{\rho_i},$$

t_{ρ_i} being the t-statistic for testing H_0: $\rho_i = 1$ in (C.1). It is shown that, given N, as $T \to \infty$, t_{ρ_i} weakly converges to $t_{iT} = \dfrac{\int_0^1 W_{iz} dW_{iz}}{\sqrt{\int_0^1 W_{iz}^2}}$, where W_{iz} denotes a Brownian motion[134]. Assuming t_{iT} 's to be independent and identically distributed with finite mean and variance, the *IPS* test statistic is derived as

$$t_{IPS} = \frac{\sqrt{N}(\bar{t} - E(t_{iT}; H_0 : \rho_i = 1)}{\sqrt{\text{var}(t_{iT}; H_0 : \rho_i = 1)}}. \tag{C.2}$$

So far as the actual test procedure is concerned, *IPS* provide table of estimates of $E(t_{iT}; H_0 : \rho_i = 1\forall i)$ and corresponding $\text{var}(t_{iT}; H_0 : \rho_i = 1\forall i)$ for different values of T and p computed by stochastic simulation for two versions of the *ADF(p)* regression–viz., $\Delta y_t = \alpha + \beta y_{t-1} + \sum_{j=1}^{p} \gamma_j \Delta y_{t-j} + error$ for the *without time trend case* and $\Delta y_t = \alpha + \delta t + \beta y_{t-1} + \sum_{j=1}^{p} \gamma_j \Delta y_{t-j} + error$ for the *with time trend case*. Given these and the computed value of $\bar{t}$ for the given panel data, t_{IPS} is calculated using (C.2). The table of corresponding critical values for the given values of N and T and various levels of significance are provided in Im et al (2003).

C.2. Co-integration Test for Panel Data

Given a set of panel data on *(K+1)* variables $y, x_j, j = 1, K$, the single equation *IPS* cointegration test proceeds as follows: First, the linear regression equation $y_{it} = \sum_{j=1}^{K} \beta_{ji} x_{jit} + error$ is estimated separately for *i =1, 2,..., N* individual units and the regression residuals

$$e_{it} = y_{it} - \sum_{j=1}^{K} \hat{\beta}_{ji} x_{jit}, i = 1,2,..., N; t = 1,2,..., T \tag{C.3}$$

[134] Brownian motion is also called *Wiener Process* (see, Hamilton (1994), ch-17, p-478).

are obtained, where $\hat{\beta}_{ji}$'s denote the estimated parameters of the regression equation for the *ith* unit. These estimated linear regression equations may be taken as estimate of the long run equilibrium relationship between *y* and the *x*'s, in case the variables turn out to be cointegrated[135]. Next, for each *i* the following *ADF(p)* equation is estimated:

$$e_{it} = \lambda e_{it-1} + \sum_{j=1}^{p} \theta_{ij} \Delta e_{it-j} + z'_{it}\gamma + v_{itp} \tag{C.4}$$

Where $z'_{it}\gamma$ is same as defined for equation (C.1) above and v_{itp} is the equation disturbance term assumed to be a white noise. Here also one may consider two alternative specifications of equation (C.4) - viz., one without a time trend and another with a time trend. The *IPS* methodology of cointegration[136] test for the set of variables under consideration thus involves the test of unit root for the regression residuals $\{e_{it}\}$- i.e., the null hypothesis H_0: $\lambda = 1$ (i.e., no cointegration) is tested against the alternative hypothesis H_1: $\lambda < 1$ (i.e., cointegration). In our empirical exercise, we have performed the cointegration test twice, viz., once treating logarithm of *PCGDP* (i.e., *y*) as the dependent variable and logarithm of *PCCO2* (i.e., *x*) as the independent variable and again reversing the status of these variables.

C.3. Estimation of ECM from Panel Data

Once the pair of variables (x, y) has been found to be cointegrated, the next step in the Engle – Granger methodology is to model the short run variations of the variables. This is done by estimating the *ECM*. For a bivariate case as the present one, the *ECM*, which is implied by the well known *Granger Representation Theorem* (see Hamilton (1994), Ch.19, pp. 581-582), is expressed as either of the following linear regression equations:

$$\Delta y_{it} = \mu_{yx} + \sum_{j=1}^{T_{11}} \alpha_{1j} \Delta y_{it-j} + \sum_{j=1}^{T_{12}} \beta_{1j} \Delta x_{it-j} + \eta_{yx} ECY_{it-1} + u_{1it} \tag{C.5}$$

$$\Delta x_{it} = \mu_{xy} + \sum_{j=1}^{T_{21}} \beta_{2j} \Delta x_{it-j} + \sum_{j=1}^{T_{22}} \alpha_{2j} \Delta y_{it-j} + \eta_{xy} ECX_{it-1} + u_{2it}\,. \tag{C.6}$$

Here Δ denotes the difference operator; $T_{lm}, l, m = 1,2$ denotes the number of lagged values of Δy_i and Δx_i that affect the current value of these *differenced* variables; μ, α, β and η denote regression parameters; $u_{lit}, l = 1,2$ are the equation disturbance terms (that should be white noises when the *ECM* has been adequately specified); and finally, $ECY_{it} = y_{it} - \hat{\phi}_0 - \hat{\phi}_1 x_{it}$ and $ECX_{it} = x_{it} - \hat{\varphi}_0 - \hat{\varphi}_1 y_{it}$ are the error correction terms

---

[135] It may be noted that when the variables are cointegrated, the true relationship underlying this linear regression equation is a long run equilibrium relationship between y and the x's. Kao, Chiang and Chen (1999) pointed out that for a set of cointegrated variables the use of *OLS* to estimate this long run equilibrium relationship from the given set of panel data will give biased results in a finite sample and recommended the use of Dynamic *OLS (DOLS)* for minimisation of such bias. See Kao and Chiang (2000) for the definition of *DOLS*.

[136] Panel data cointegration test is also performed by Kao (1999), McCoskey and Kao (1998).

(hereafter refereed to as *EC terms*) measuring deviation of $y_{it}(x_{it})$ from the corresponding long run equilibrium value, given $x_{it}(y_{it})$.[137] The parameters η_{yx} and η_{xy} in equations (C.5) and (C.6) are called the adjustment parameters. They are expected to have negative values[138]. In this set up the nature of Granger causality is determined as follows:

(1) if $\beta_{1j} = 0$ for all *j* and $\eta_{yx} = 0$, *x* may be said not to *Granger cause y*;
(2) if $\alpha_{2j} = 0$ for all *j* and $\eta_{xy} = 0$, *y* may be said not to *Granger cause x*;
(3) if (1) holds but (2) does not, *Granger causality* may be said to be *unidirectional from y to x*;
(4) Conversely, if (1) does not hold but (2) does, *Granger causality* may be said to be *unidirectional from x to y*;
(5) if both (1) and (2) do not hold, *Granger causality* between *x* and *y* may be said to be *bi- directional*; and finally
(6) if both (1) and (2) hold, *Granger causality* between *x* and *y* may be said to be absent (see Enders (1995), Glasure and Lee (1997) and Asafu-Adjaye (2000) for details).

In the present exercise, equations (C.5) and (C.6) (henceforth referred to as model I and model II, respectively) were estimated separately for each country group, using the panel data set for the country group. Country group-specific inference about the nature of *Granger causality* between *x* and *y* were then drawn by performing appropriate test of hypothesis for the relevant parameters of model I and II, as laid down above. For example, to test the null hypothesis that *x* does not *Granger cause y,* one should perform an F-test for the null

[137] Note that here $y_{it} = \phi_0 + \phi_1 x_{it} + \varepsilon_{1it}$ and $x_{it} = \varphi_0 + \varphi_1 y_{it} + \varepsilon_{2it}$ are alternative representations of the (population) long run *equilibrium* relationship between *y* and *x*, where ε 's are the stationary error terms. As y and *x* are cointegrated, by the definition of cointegration for some constants, $\omega_0 + \omega_1 y_{it} + \omega_2 x_{it} = \varepsilon_{it}$, where ε_{it} is a stationary error term and $\omega = (\omega_0, \omega_1, \omega_2)$ is the non- normalized cointegrating vector. Thus, by normalizing ω one may write the long run equilibrium relationship for *(y,x)* in either form as shown above.

[138] This is for the following reason. If, for example, $ECY_{it-1} > 0$ for some *i,t,* it means that the realized value of y_i exceeded the corresponding long run equilibrium level at *t-1*, given x_{it}. Now since y_i and x_i are cointegrated, once a positive deviation from the long run equilibrium level takes place, the actual value must try to move in the opposite direction in subsequent time points in an attempt to restore the long run equilibrium and hence the negative sign of η_{yx} and η_{xy}.

hypothesis $H_0 : \beta_{1j} = 0, j = 1,2,...,T_{12}, \eta_{yx} = 0$, using model I. Similarly, to test the null hypothesis that y does not *Granger cause* x, an F-test for the null hypothesis $H_0 : \alpha_{2j} = 0, j = 1,2,...,T_{22}, \eta_{xy} = 0$ using model II will be required. Given the results of these two basic F-tests, the remaining null hypotheses (3)- (6) laid down above can be tested.

REFERENCES

Aghion, P. and P. Howitt, (1998), Endogenous Growth Theory. The MIT Press, Cambridge, M.A.

Agras, J. and D. Chapman (1999): "A dynamic approach to the Environmental Kuznets Curve hypothesis", *Ecological Economics* 28(2), 267 - 277.

Andreoni, J. and A. Levinson (2001): "The simple analytics of the environmental Kuznets curve", *Journal of Public Economics* 80(2), 269 - 286.

Anselin, L. (1988): *Spatial Econometrics: Methods and Models*, Kulwer Academic Publications, Dordrecht.

Antweiler, W., B. R. Copeland and M. S. Taylor (2001): "Is Free Trade Good for the Environment?", *American Economic Review* 91(4), 877 – 908.

Arrow, K., B. Bolin, R. Costanza, P. Dasgupta, C. Folke, C. S. Holling, B.O. Janson, S. Levin, K.G. Maler, C. Perrings, and D. Pimental (1995): "Economic growth, carrying capacity, and the environment", *Ecological Economics* 15, 91-95.

Asafu-Adjaye, J. (2000): "The relationship between energy consumption, energy price and economic growth: time series evidence from Asian developing countries", *Energy Economics* 22, 615 - 625.

Aitchison, J. and J. A. C. Brown (1957): *Lognormal distribution with special reference to its users in economics*, Cambridge University Press, Cambridge.

Baltagi, B. H. (1999): *Econometric Analysis of Panel Data,* John Wiley & Sons, New York.

Barro, R. J. and X. Sala-I-Martin, (1995): *Economic Growth*, McGraw-Hill, Inc., New York.

Basu, S., and Weil, D. N., 1998, "Appropriate Technology and Growth", *Quarterly Journal of Economics*, Vol.-113, No.4, 1025 – 1054.

Beckerman, W. (1992): "Economic growth and the environment: Whose growth? Whose environment?", *World Development* 20, 481 - 496.

Becker, R. A., (1982), "Intergenerational Equity: The Capital-Environment Trade-off", *Journal of Environmental Economics and Management*, vol-9, 165-185.

Beltratti, A., (1996), *Models of Economic Growth with Environmental Assets*, Kluwer Academic Publishers, Dordrecht.

Berrens, R., A. Bohara, K. Gawande and P. Wang (1997): "Testing the inverted U - hypothesis for U.S. hazardous waste: An application of the generalized gamma model", *Economic Letters* 55(3), 435 - 440.

Bertinelli, L. and Strobl, E., 2005, "The Environmental Kuznets Curve semi-parametrically revisited", *Economics Letters*, vol.-88, 350 – 357.

Bhattarai, M. and M. Hammig (2001): "Institutions and the Environmental Kuznets Curve for Deforestation: A Crosscountry Analysis for Latin America, Africa, and Asia", *World Development* 29(6), 995-1010.

Bimonte, S. (2002): "Information access, income distribution, and the Environmental Kuznets Curve", *Ecological Economics* 41, 145 - 156.

Bommer, Rolf (1999): "Environmental Policy and Industrial Competitiveness: The Pollution Haven Hypothesis Reconsidered", *Review of International Economics* 7(2), 342 – 355.

Bovenberg, A. L. and S. Smulders, (1995), "Environmental quality and pollution-augmenting technological change in a two-sector endogenous growth model", *Journal of Public Economics*, vol.-57, 369-391.

Bradford, D. F., R. Schlieckert, and S. H. Shore (2000): "The Environmental Kuznets Curve: Exploring a fresh specification", *NBER working paper* 8001.

Breitung, J. and M. H. Pesaran (2005): "Unit Roots and Cointegration in Panels", CESifo *Working Paper* No. 1565.

Bresnahan, T. F. and Trajtenberg, M., 1995, "General Purpose Technologies: 'Engines of Growth'?", *Journal of Econometrics*, vol.-65(1), 83-108.

Brock, W. A. and M. S. Taylor (2004a): "The Green Solow Model", *NBER Working paper* 10557.

Brock, W. A. and M. S. Taylor (2004b): "Economic Growth and the Environment: A Review of Theory and Empirics", *NBER Working Paper* 10854.

Bulte, E.H. and D. P. van Soest (2001): "Environmental degradation in developing countries: households and the (reverse) Environmental Kuznets Curve", *Journal of Development Economics* 65, 225 – 235.

Canas, A., P. Ferrao, and P. Conceicao (2003): "A new environmental Kuznets curve? Relationship between direct material input and income per capita: evidence from industrialised countries", *Ecological Economics* 46, 217 –229.

Carson, R. T., Y. Jeon, and D. R. McCubbin (1997): "The relationship between air pollution emissions and income: US data", *Environment and developmental Economics* 2, 433 - 450.

Cass, D., (1965), "Optimum growth in an Aggregative Model of capital accumulation", *Review of Economics studies*, vol.-32, 233 – 240.

Cassou, S. P., and Hamilton, S. F., 2004, "The transition from dirty to clean industries: optimal fiscal policy and the environmental Kuznets curve", *Journal of Environmental Economics and Management,* vol.- 48, 1050–1077.

Cavlovic, T. A., K.H. Baker, R. P. Berrens and K. Gawande (2000): "A Meta-analysis of Environmental Kuznets Curve studies", *Agricultural and Resource Economics Review* 29(1), 32 - 42.

Chaudhuri, S., and A. Pfaff (1998): "Household income, Fuel Choice, and Indoor Air Quality: Microfoundations of an Environmental Kuznets Curve", Economics department, Columbia University, *mimeo.*

Cheng, B. S. (1999): "Causality between energy consumption and economic growth in India: An application of co-integration and error-correction modeling", *Indian Economic Review* 34, 39 - 49.

Cheng, B.S. (1996): "An Invetigation of Co-integration and Causality between energy consumption and Economic Growth", *The Journal of Energy and Development* 21(1), 73 – 82.

Cheng, B. S. and T. W. Lai (1997): "An investigation of cointegration and causality between energy consumption and economic activity in Taiwan", *Energy Economics* 19, 435 - 444.

Chimeli, A. B., and Braden, J. B., 2005, "Total factor productivity and the environmental Kuznets curve", *Journal of Environmental Economics and Management*, vol.- 49, 366–380.

Cole, M. A. (2004): "Trade, the pollution haven hypothesis and the environmental Kuznets curve: examining the linkages", *Ecological Economics* 48 (1), 71-81.

Cole, M. A. (2003): "Development, trade, and the environment: how robust is the Environmental Kuznets Curve?", *Environment and Development Economics* 8, 557 –580.

Cole, M. A., A. J. Rayner, and J. M. Bates (1997): "The Environmental Kuznets Curve: An empirical analysis", *Environment and Development Economics* 2, 401 - 416.

Coondoo, D. and S. Dinda (2002): "Causality between income and emission: a country group-specific econometric analysis", *Ecological Economics* 40 (3), 351 - 367.

Copeland, B. R., and M. S. Taylor (2004): "Trade, Growth and the Environment", *Journal of Economic Literature* XLII, 7 – 71.

Copeland, B. R., and M. S. Taylor (1995): "Trade and environment: a partial synthesis", *American Journal of Agricultural Economics* 77, 765 - 771.

D'Arge, R.C. and K.C. Kogiku, (1973), "Economic growth and the environment", *Review of Economics Studies*, vol.-40, 61-77.

Dasgupta, S., B. Laplante, H. Wang, and D. Wheeler (2002): "Confronting the Environmental Kuznets Curve", *Journal of Economic Perspectives* 16(1), 147–168.

Dasgupta, S., A. Mody, S. Roy, and D. Wheeler (2001): "Environmental Regulation and Development: A Cross-Country Empirical Analysis", *Oxford Development Studies* 29, 173 - 187.

de bruyn, S. M., J. C. J. M. van den bergh, and J. B. Opschoor (1998): "Economic growth and emissions: reconsidering the empirical basis of Environmental Kuznets Curves", *Ecological Economics* 25, 161 - 175.

de bruyn, S. M. and R. J. Heintz (1998): "The Environmental Kuznets Curve hypothesis", in *Handbook of Environmental Economics*, 656 – 677.

de bruyn, S. M., and J. B. Opschoor (1997): "Developments in the throughput -income relationship: theoretical and empirical observations", *Ecological Economics* 20, 255 - 268.

de bruyn, S. M. (1997): "Explaining the Environmental Kuznets Curve: structural change and international agreements in reducing sulphur emissions", *Environment and Development Economics* 2, 485 – 503.

Dinda, S., 2007, "Technological Progress towards Sustainable Development", *International Journal of Global Environmental Issues*, (forth coming).

Dinda, S. (2005): "A theoretical basis for the Environmental Kuznets Curve", *Ecological Economics* 53(3), 403-413.

Dinda, S. (2004): "Environmental Kuznets Curve Hypothesis: A Survey", *Ecological Economics* 49(4), 431-455.

Dinda, S., and D. Coondoo (2006): "Income and Emission: A Panel Data based Cointegration Analysis", *Ecological Economics* 57(2), 167 - 181.

Dinda, S., D. Coondoo, and M. Pal (2000): "Air quality and economic growth: an empirical study", *Ecological Economics* 34(3), 409 - 423.

Elbasha, E. H. and T. L. Roe, (1996), "On Endogenous Growth: The Implications of Environmental Externalities", *Journal of Environmental Economics and Management*, vol-31, 240-268.

Ekins, P. (1998): Economic growth and Environmental sustainability, ch.7, 182–214.

Engle, R.F. and C.W.J. Granger (1987): "Cointegration and error correction representation, estimation, and testing", *Econometrica* 55(2), 251 - 276.

Enders, W. (1995): Applied Econometric Time Series, John Wiley & Sons, Inc., USA.

Ezzati, M., B. H. Singer, and D. M. Kammen (2001): "Towards an Integrated Framework for Development and Environmental Policy: The Dynamics of Environmental Kuznets Curves", *World Development* 29(8), 1421 - 1434.

Forster, B.A., (1973), "Optimal capital accumulation in a polluted environment", *Southern Economic Journal*, vol.-39, 544-547.

Gangadharan, L. and M. R. Valenzuela (2001): "Interrelationships between income, health and the environment: extending the Environmental Kuznets Curve hypothesis", *Ecological Economics* 36(3), 513 - 531.

Gawande, K., R. P. Berrens, and A.K. Bohara (2001): "A consumption-based theory of the environmental Kuznets curve", *Ecological economics* 37(1), 101 - 112.

Gawande, K., A. K. Bohara, R. P. Berrens and P. Wang (2000): "Internal migration and the environmental Kuznets curve for U.S. hazardous waste sites", *Ecological Economics* 33(1), 151 - 166.

Gengenbach, C., F. C. Palm and J. P. Urbain (2004): "Panel Unit Root Tests in the Presence of Cross- Sectional Dependencies: Comparision and Implications for Modelling", Department of Quantitative Economics, University of Maastricht, *Research Memorandum* 040.

Gengenbach, C., F. C. Palm and J. P. Urbain (2005): "Panel Cointegration Testing in the presence of Common Factors", Department of Quantitative Economics, University of Maastricht, *Research Memorandum* 050.

Glasure, Yong U. and Lee, Aie-Rie (1997): "Co-integration, error-correction, and the relationship between GDP and energy: The case of South Korea and Singapore", *Resource and Energy Economics* 20, 17 – 25.

Goulder, L. H. and S. H. Schneider (1999): "Induced technological change and the attractiveness of CO_2 abatement policies", *Resource and Energy Economics* 21, 211-253.

Gradus, R. and S. Smulders, (1993), "The Trade-off Between Environmental Care and Long-term Growth –Pollution in Three Prototype Growth Models", *Journal of Economics*, vol.-58(1), 25-51.

Granger, C.W.J. (1969): "Investigating Causal Relations by Econometric models and cross-spectral methods", *Econometrica* 37(3), 424 – 438.

Grossman, G. M. and A. B. Krueger (1995): "Economic growth and the Environment", *Quarterly Journal of Economics* 110(2), 353 - 377.

Grossman, G. M. and A. B. Krueger (1993): "Environmental impacts of the North American Free Trade Agreement", in P. Garber, (ed.) The U.S.-Mexico Free Trade Agreement., Cambridge: MIT Press, 13 –56.

Grossman, G. M. and A. B. Krueger (1991): "Environmental impacts of the North American Free Trade Agreement", *NBER working paper* 3914.

Gruver, G.W., (1976), "Optimal investment in pollution control in a neoclassical growth context", *Journal of Environmental Economics and Management*, vol-3, 165-177.

Halkos, George E. (2003): "Environmental Kuznets Curve for sulfur: evidence using GMM estimation and random coefficient panel data model", *Environment and Development Economics* 8, 581 – 601.

Hamilton, J. D. (1994): *Time Series Analysis,* Princeton University Press, New Jersey, USA.

Harbaugh, W., A. Levinson and D. M. Wilson (2002): "Re-examining the Empirical Evidence for an Environmental Kuznets Curve", *The Review of Economics and Statistics* 84(3), 541-551.

Heerink, N., A. Mulatu and E. Bulte (2001): "Income inequality and the environment: aggregation bias in environmental Kuznet curves", *Ecological Economics* 38, 359 – 367.

Hettige, H., R. E. B. Lucas and D. Wheeler (1992): "The Toxic Intensity of Industrial Production: Global patterns, Trends and Trade Policy", *American Economic Review* 82, 478 - 481.

Hettige, H., M. Mani and D. Wheeler (2000): "Industrial pollution in economic development: the environmental Kuznets curve revisited", *Journal of Development Economics* 62, 445 - 476.

Hilton, F. G. H. and A. Levinson (1998): "Factoring the Environmental Kuznets Curve: Evidence from Automotive Lead Emissions", *Journal of Environmental Economics and Management* 35, 126 - 141.

Holtz-Eakin, D. and T. M. Selden (1995): "Stoking the fires?: CO_2 emissions and economic growth", *Journal of Public Economics* 57, 85 - 101.

Horvath, R. J. (1997): "Energy consumption and the Environmental Kuznets Curve Debate", Department of Geography, University of Sydney, Australia, *mimeo.*

Hsiao, C., M. (1986): Analysis of Panel data, Cambridge University Press, Cambridge.

Im, K. S., M. H. Pesaran and Y. Shin (2003): "Testing for unit roots in heterogeneous panels", *Journal of Econometrics* 115, 53-74.

Jensen, J. and T. N. Rasmussen (2000): "Allocation of CO_2 emission permits: a general equilibrium analysis of policy instruments", *Journal of Environmental Economics and Management* 40, 111-136.

Jha, R. and K.V. Bhanu Murthy (2003): "An inverse global environmental Kuznets curve", *Journal of Comparative Economics* 31, 352 – 368.

Jha, S. K. (1996): "The Kuznets Curve: A Reassessment", *World Development* 24(4), 773 - 780.

John, A. and R. Pecchenino (1994): "An overlapping generations model of growth and the environment", *Economic Journal* 104, 1393 - 1410.

Judge, G., K. C. Hill, W.G. Griffiths, H. Lutkepohl and T.C. Lee (1985): The Theory and Practice of Econometrics, 2nd Edn, John Wiley & Sons, Inc, New York.

Kahn, M.E. (1998): "A household level environmental Kuznets curve", *Economics Letters* 59, 269 - 273.

Kakwani, N. C. (1980): *Income Inequality and Poverty: methods of estimation and policy application*, Oxford University Press, Calcutta.

Kao, C. (1999): "Spurious regression and residual-based tests for cointegration in panel data", *Journal of Econometrics*, vol.-90, 1 – 44.

Kao, C. and M. H. Chiang (2000): "On the Estimation and Inference of a Cointegrated Regression in Panel Data", *Advances in Econometrics* 15, 179-222.

Kao, C., M. Chiang and B. Chen (1999): "International R&D spillovers: An application of estimation and inference in panel cointegration", *Oxford Bulletin of Economics and Statistics* 61(4), 693 – 711.

Kaufmann, R. K., B. Davidsdottir, S. Garnham and P. Pauly (1998): "The determinants of atmospheric SO_2 concentrations: reconsidering the environmental Kuznets curve", *Ecological Economics* 25, 209 - 220.

Keeler, E., M. Spence and R. Zeckhauser, (1972), "The optimal control of pollution", *Journal of Economic Theory*, vol.-4, 19-34.

Khanna, N., 2002, "The income elasticity of non-point source air pollutants: revisiting the environmental Kuznets curve", *Economics Letters*, Vol.-77, 387 – 392.

Komen, R., S. Gerking and H. Folmer (1997): "Income and Environmental R & D: empirical evidence from OECD countries", *Environment and Development Economics* 2, 505 - 515.

Koop, G. and L. Tole (1999): "Is there an environmental Kuznets curve for deforestation?" *Journal of Development Economics* 58, 231 - 244.

Krautkraemer, J. A., (1985), "Optimal growth, Resource Amenities and the preservation of Natural Environments", *Review of Economic Studies*, vol.-LII, 153-170.

Kristrom, B. and P. Riera (1996): "Is the income elasticity of environmental improvements less than one?", *Environmental and Resource Economics* 7, 45-55.

Kuznets, Simon (1955): "Economic Growth and income inequality", *American Economic Review* 45, 1 - 28.

Levin, A. and C. F. Lin (1993): "Unit Root Tests in Panel Data: New Results", University of California, San Diego, *Working paper* 56-93.

Liang, F. H., (2006): "Does Foreign Direct Investment Harm the Host Country's Environment? Evidence from China", UC Berkeley, *mimeo*.

Liddle, B. (2001): "Free trade and the environment-development system", *Ecological Economics* 39, 21 – 36.

Lieb, C. M. (2002): "The Environmental Kuznets Curve and satiation: a simple static model", *Environment and Development Economics* 7, 429 - 448.

Lindmark, M. (2002): "An EKC-pattern in historical perspective: carbon dioxide emissions, technology, fuel prices and growth in Sweden 1870 – 1997", *Ecological Economics* 42, 333 - 347.

List, J. A. and C. A. Gallet (1999): "The environmrntal Kuznets curve: does one size fit all?", *Ecological Economics* 31, 409 - 423.

Liu, X., 2005, "Explaining the relationship between CO2 emissions and national income—the role of energy consumption", *Economics Letters*, Vol.-87, 325 – 328.

Lopez, R. and S. Mitra (2000): "Corruption, Pollution, and the Kuzents Environment Curve", *Journal of Environmental Economics and Management* 40(2), 137 - 150.

Lopez, R. (1994): "The environment as a factor of production: the effects of economic growth and trade liberalization", *Journal of Environmental Economics and management* 27, 163 - 184.

Low, P. and A. Yeats (1992): "Do 'dirty' industries migrate?" In P. Low (ed.), *International Trade and environment*., World Bank., Washington, D.C.

Lucas, R. E. B., D. Wheeler and H. Hettige (1992): "Economic development, Environmental Regulation and the international migration of toxic industrial pollution: 1960 – 1988", in P. Low (ed.) *International Trade and the Environment*, World Bank discussion paper 159, World Bank., Washington, D.C.

Maddala, G. S. and In-Moo Kim (1999): Unit roots, Co-integration and Structural change, Cambridge University Press, Cambridge, U.K.

Maddison, D., 2006, "Environmental Kuznets curves: A spatial econometric approach", *Journal of Environmental Economics and Management*, vol.-51, 218–230.

Magnani, E. (2000): "The Environmental Kuznets Curve, environmental protection policy and income distribution", *Ecological Economics* 32, 431 - 443.

Mani, M. and D. Wheeler (1998): "In search of pollution havens? Dirty industry in the world economy: 1960 – 1995"., *Journal of Environment and Development* 7(3), 215 - 247.

McConnell, K.E. (1997): "Income and the demand for environmental quality", *Environment and Development Economics* 2, 383 - 399.

McCoskey, S. and C. Kao (1998): "A residual-based test of the null of cointegration in panel data", *Econometric Reviews* 17, 57-84.

Meadows, D. H., D. L. Meadows, J. Randers and W. Behrens (1972): The Limits to Growth, Universe Books, New York.

Michel, P. and Rotillon (1995). "Disutility of Pollution and Endogenous Growth". *Environmental and Resource Economics,* vol.-6, 279-300.

Mohr, R. D., (2002), "Technical Change, External Economies, and the Porter Hypothesis", *Journal of Environmental Economics and Management*, vol.- 43, 158 -168.

Moomaw, W. R. and G. C. Unruh (1997): "Are Environmental Kuznets Curve misleading us? The case of CO_2 emissions", *Environment and Development Economics* 2, 451 – 463.

Mukhopadhyay, K., Chakraborty, D. and Dietzenbacher, E., (2005): "Pollution Haven and Factor Endowment Hypotheses Revisited: Evidence from India", *paper presented in the Fifth International Input-Output Conference at Renmin University in Beijing, China, during June 27 – July 1.*

Mulligan, C. B. and X. Sala-I-Martin, (1993), "Transitional Dynamics in Two-sector Models of Endogenous Growth", *Quarterly Journal of Economics*, vol.-108(3), 739-773.

Munasinghe, M. (1999): "Is environmental degradation an inevitable consequence of economic growth: tunneling through the environmental Kuznets curve", *Ecological Economics* 29, 89 - 109.

Oak Ridge National Laboratory, CDIAC, Environmental Science Division, 1998, Estimates of global, regional and national CO_2 emissions from fossil fuel burning, cement manufacturing and gas flaring: 1755 – 1996, available at http://www.cdiac.esd.ornl.gov/epubs/ndp030/global97.ems, (updated 2000).

Panayotou, T. (1999): "The economics of environments in transition", *Environment and Development Economics* 4(4), 401 - 412.

Panayotou, T. (1997): "Demystifying the environmental Kuznets curve: turning a black box into a policy tool", *Environment and Development Economics* 2, 465-484.

Panayotou, T. (1993): "Empirical tests and policy analysis of environmental degradation at different stages of economic development", ILO, *Technology and Employment Programme*, Geneva.

Parente S. L. and E. C. Prescott (2000) *Barriers to Riches: The Walras-Pareto Lectures*, MIT Press, Cambridge MA.

Pargal, S. and D. Wheeler (1996): "Informal regulation of Industrial Pollution in Developing countries: Evidence from Indonesia", *Journal of Political Economy* 104 (6), 1314 - 1327.

Pasche, Markus (2002): "Technical progress, structural change, and the environmental Kuznets curve", *Ecological Economics* 42, 381-389.

Perman, R. and David I. Stern (2003): "Evidence from Panel Unit root and Cointegration tests that the Environmental Kuznets Curve does not exist", *Australian Journal of Agricultural and Resource Economics* 47, 325 – 347.

Popp, D., 2004, "ENTICE: endogenous technological change in the DICE model of global warming", *Journal of Environmental Economics and Management*, vol.- 48, 742–768.

Quah, D. (1994): "Exploiting cross-section variation for unit root inference in dynamic data", *Economics Letters* 44, 9 – 19.

Ravallion, Martin., M. Heil and J. Jalan (2000): "Carbon emissions and income inequality", *Oxford Economic Papers* 52, 651 – 669.

Reis, A. B., (2001), "Endogenous Growth and the Possibility of Eliminating Pollution", *Journal of Environmental Economics and Management*, vol.- 42, 360 - 373.

Robers, J. T. and P. K. Grimes (1997): "Carbon intensity and economic development 1962 - 1991: A brief exploration of the environmental Kuznets curve", *World development* 25(2), 191 - 198.

Roca, J. (2003): "Do individual preferences explain Environmental Kuznets curve?", *Ecological Economics* 45(1), 3 - 10.

Rothman, D. S. (1998): "Environmental Kuznets curve - real progress or passing the buck?: A case for consumption-base approaches", *Ecological Economics* 25, 177-194.

Schmalensee, R., T. M. Stoker and R. A. Judson (1998): "World Carbon Dioxide Emissions: 1950- 2050", *The Review of Economics and Statistics* 80(1),15-27.

Scruggs, L. A. (1998): "Political and economic inequality and the environment", *Ecological Economics* 26, 259 –275.

Selden, T. M., A. S. Forrest and J.E. Lockhart (1999): "Analyzing the reductions in US air pollution emissions: 1970 – 1999", *Land Economics* 75(1), 1 - 21.

Selden, T. M. and D. Song (1995): "Neoclassical growth, the J Curve for abatement, and the inverted -U Curve for pollution", *Journal of Environmental Economics and management* 29, 162 - 168.

Selden, T. M. and D. Song (1994): "Environmental Quality and Development: Is There a Kuznets Curve for Air Pollution Emissions?", *Journal of Environmental Economics and management* 27, 147 - 162.

Sengupta, R. P. (1997): "CO_2 emission - income relationship: policy approach for climate control", *Pacific Asia Journal of Energy* 7(2), 207 - 229.

Shafik, N. (1994): "Economic development and environmental quality: an econometric analysis", *Oxford Economic Papers* 46, 757 - 773.

Shukla, V. and K. Parikh (1992): "The environmental consequences of urban growth: cross - national perspectives on economic development, air pollution, and city size", *Urban geography* 12, 422 - 449.

Smulder, S. and Bretschger, L., (2000), "Explaining Environmental Kuznets Curves: How Pollution Induces Policy and New Technologies"; W. Ouestlati and G. Rotillon (eds.): *Macroeconomics perspectives on Environmental Concerns*, Edward Elgar.

Smulder, S., Bretschger, L., and Egli, H., (2005), "Economic Growth and the Diffusion of Clean Technologies: Explaining Environmental Kuznets Curves"; *Working Paper* 05/42, ETH Zurich.

Stephens, J.K. (1976), "A Relatively optimistic Analysis of Growth and Pollution in a Neoclassical framework", *Journal of Environmental Economics and Management,* vol.-3, 85-96.

Stern, David I. (2004): "The Rise and Fall of the Environmental Kuznets Curve", *World Development* 32 (8), 1419 1439.

Stern, David I. (2002): "Explaining changes in global sulfur emissions: an econometric decomposition approach", *Ecological Economics* 42, 201 –220.

Stern, David I. (2000): "A multivariate cointegration analysis of the role of energy in the US macroeconomy", *Energy Economics* 22, 267 - 283.

Stern, David I. (1998): "Progress on the Environmental Kuznets Curve?", *Environment and Development Economics* 3, 175 - 198.

Stern, David I. (1993): "Energy and economic growth in the USA, a multivariate approach", *Energy Economics* 15, 137 – 150.

Stern, David I. and M. S. Common (2001): "Is There an Environmental Kuznets Curve for Sulfur?", *Journal of Environmental Economics and Management* 41(2), 162-178.

Stern, David I., M. S. Common and E. B. Barbier (1996): "Economic growth and environmental degradation: The environmental Kuznets curve and sustainable development", *World Development* 24, 1151 - 1160.

Stokey, N. L. (1998): "Are there limits to Growth?", *International Economic Review* 39(1), 1 - 31.

Summers, R. and A. Heston (1994): "Penn World Table (Version 5.6): An Expanded Set of International Comparisons:1950–1992", NBER, PWT5.6, available at *http://www. nber.org/pwt5.6.*

Suri, V. and D. Chapman (1998): "Economic growth, trade and the energy: implications for the environmental Kuznets curve", *Ecological Economics* 25, 195 - 208.

Tahvonen, O. and J. Kuuluvainen (1993). Economic Growth, Pollution, and Renewable Resources. *Journal of Environmental Economics and Management,* vol.-24, 101—118.

Tarui, N., and Polasky, S., (2005), "Environmental regulation with technology adoption, learning and strategic behaviour", *Journal of Environmental Economics and Management*, vol.- 50, 447–467

Taskin, F. and O. Zaim (2000): "Searching for a Kuznets curve in environmental efficiency using kernel estimation", *Economics Letters* 68, 217 - 223.

Temurshoev, U., (2006): "Pollution Haven Hypothesis or Factor Endowment Hypothesis: Theory and Empirical Examination for the US and China", CERGE-EI, *Working paper 292.*

Tisdell, C. (2001): "Globalisation and sustainability: environmental Kuznets curve and the WTO", *Ecological Economics* 39, 185 -196.

Torras, M. and J. K. Boyce (1998): "Income, inequality, and pollution: a reassessment of the environmental Kuznets Curve," *Ecological Economics* 25, 147-160.

Tucker, M. (1995): "Carbon dioxide emissions and global GDP", *Ecological Economics* 15, 215 – 223.

Unruh, G. C. and W. R. Moomaw (1998): "An alternative analysis of apparent EKC -type transitions", *Ecological Economics* 25, 221 - 229.

Vincent, J. R. (1997): "Testing for environmental Kuznets curves within a developing country", *Environment and Development Economics* 2, 417 - 431.

Wheeler, D. (2000): "Racing to the Bottom? Foreign Investment and Air Pollution in Developing Countries", World Bank Development Research Group *Working Paper* No. 2524.

Withagen, C. (1995). "Pollution, Abatement and Balanced Growth". *Environmental and Resource Economics,* vol.-5, 1-8.

World Bank (1992): *World Development Report 1992*, Oxford University Press, New York.

World Bank (1986): *World Development Report 1986*, Oxford University Press, New York.

World Resource Institute, World Bank, (1998): *World Resources* 1998-99, Oxford University Press, New York.

Yang, Hao-Yen (2000): "A note on the causal relationship between energy and GDP in Taiwan", *Energy Economics* 22, 309 – 317.

INDEX

D

E

F

G

N

O

P

Q

R

S

T

U

V

W

Y

Z